Visions for a Sustainable Energy Future

Visions for a Sustainable Energy Future

By Mark A. Gabriel

LONDON AND NEW YORK

Published 2020 by River Publishers
River Publishers
Alsbjergvej 10, 9260 Gistrup, Denmark
www.riverpublishers.com

Distributed exclusively by Routledge
4 Park Square, Milton Park, Abingdon, Oxon OX14 4RN
605 Third Avenue, New York, NY 10158

First published in paperback 2024

Library of Congress Cataloging-in-Publication Data

Gabriel, Mark A., 1954-
 Visions for a sustainable energy future / by Mark A. Gabriel.
 p. cm.
 Includes bibliographical references and index.
 ISBN 0-88173-513-2 (alk. paper) -- ISBN 978-8-7702-2257-0 (electronic) -- ISBN 978-0-8493-9817-9 (Taylor & Francis distribution : alk. paper)
    1. Electric power production--Forecasting. 2. Electric utilities. 3. Energy policy. 4. Sustainable development. I. Title.

TK1005.G24 2008
333.793'2--dc22

2007045383

Visions for a Sustainable Energy Future/by Mark A. Gabriel.
First published by Fairmont Press in 2008.

©2008 River Publishers. All rights reserved. No part of this publication may be reproduced, stored in a retrieval systems, or transmitted in any form or by any means, mechanical, photocopying, recording or otherwise, without prior written permission of the publishers.

Routledge is an imprint of the Taylor & Francis Group, an informa business

Publisher's Note
The publisher has gone to great lengths to ensure the quality of this reprint but points out that some imperfections in the original copies may be apparent.

978-0-88173-513-0 (The Fairmont Press, Inc.)
978-8-7702-2257-0 (online)

While every effort is made to provide dependable information, the publisher, authors, and editors cannot be held responsible for any errors or omissions.

ISBN: 978-0-8493-9817-9 (hbk)
ISBN: 978-87-7004-573-5 (pbk)
ISBN: 978-1-003-15128-9 (ebk)

DOI: 10.1201/9781003151289

Dedication

To my father, Dr. John A. Gabriel, who taught me to always question and explore, my mother Laura for her support, and to my daughters, the shining loves of my life, Paula and Lana, with whom I share this legacy.

Table of Contents

Acknowledgments ix
Introduction xi

Chapter 1—Megatrends and the Energy Horizon 1
Market trends vs. Megatrends 1
Current Crossroads: The Electric Sector Today 4
Energy Business Capital Expenditures
    As a Percent of Revenue 6
Inherent Conflicts in the Electricity Sector 11
Relieving the Conflict 14

Chapter 2—Incenting Innovation:
    A Competitive Regulatory Framework 19
The Regulatory Dilemma 20
Electricity Company Responses to Deregulation: Invest in the New 24
Finding the Right (and Profitable) Customer Compact 33
Turning a Megatrend into Business Opportunity 35
Competitive Regulation: The Infrastructure Investment Model 38

Chapter 3—The Destiny of Energy Business Evolution 41
Developing a Stable Business Model 44
The Art of Business Reinvention:
    A Classic and Stable Business Model 56
The Unique Attributes of the Electricity Industry 62
Succeeding with the Destiny of Business Evolution 66

Chapter 4—The Destiny of Demographics 71
The Aging Workforce—A Gap of Knowledge 74
Growing Population Shifts—A Booming Retirement Issue 91
Increased Consumer Demand—A Boomer Phenomenon 95
Succeeding with the Destiny of Demographics 103

Chapter 5—The Destiny of Carbon Constraints and
    Capacity Demands 109
Supply-side Issues and the Carbon Conflict 113

Fueling the Future: Facing the Conflict 120
Incentives to Help Replace Conflict and Costly Compliance 161
Succeeding with the Megatrend of Conflict..

Chapter 6—The Destiny of Intelligent Infrastructure 169
Our National Power Grid… Now 172
The Vision—Enhanced Power Flow with Digital Control 180
Succeeding with the Megatrend of Connectivity 195

Chapter 7—The Destiny of Customer Engagement 201
It's Your Father's Energy Interface 203
The Intelligent Energy Portal 206
The Art of Energy Efficiency and Smart Control 215
Succeeding with the Megatrend of Choice 218

Chapter 8—The Destination: A Transformed Electricity Sector............. 223
A Vision for the Industry's Future... 228
Transformative Paths in Telecommunications.. 233
Getting There from Here 236
A Healthy Balance 236
Repair, Replace, Rebuild vs. Investing in Innovation.............................. 237
Building a Model for the Future 240

Conclusion... 245
Ten Ideas for the Future 245
Index 249

Acknowledgments

This book would not have been possible without the knowledge, support and breadth of experiences given to me by my colleagues in the utility industry, the people at R.W. Beck who supported the completion of this book and the Electric Power Research Institute where I had the fantastic opportunity to be exposed to the broad array of issues and opportunities in the electric power industry. EPRI is amazing in many ways as a unique institution for the industry both in practical terms for the work it does and in philosophical as it is "allowed to dream" on behalf of the nation's most valuable and critical resource provider.

This book was made possible through the access to information and experts, the industry's openness in giving me the ability to reach across a vast landscape with unparalleled access to its leadership and the fortunate circumstances that have provided me with opportunities to test ideas. There are many people to thank in the writing of a book such as this from my colleagues at R.W. Beck to the officer corps at EPRI to my start in the industry at Central Vermont Public Service. At the risk of leaving someone out, utility industry leaders, past and present have provided context, experience and guidance. Each has helped form the constructs under which I have been able to build pieces of this book.

I need to recognize the contribution of Dave Boutacoff who was able to bring life to the technical portions of this book. Dave took my rambling about technology and made it real by bringing in the facts and not just my conjecture. While my own technical experience has been focused in the retail and power delivery arena and I am fortunate to be a mile wide in understanding technologies, most of that is but a few inches deep. David provided the backbone of technical writing support that has been critical in this effort.

Fred Potter, a former EPRI colleague and former colleague at the CFO at Central Vermont Public Service, helped to keep the business perspective in balance for me as I worked through the theories in this book. His straightforward, no-nonsense approach to business and finance, and his ability to cut through the flowery language I tend to produce helped build a stronger case.

Holly Larsen helped bring the book to its final form by adding her own special critical eye to ensure that all the pieces of this puzzle fit. Her fresh perspective as the project drew to a close was a critical contribution to the final product.

Most importantly, this book would not have been possible without the overall management, support, guidance, prodding, editing and continual questioning of Ann Iverson. Ann's management, focus, organization, and ability to keep me on track and translate my theories into book form made this all possible.

Introduction

My father, a college professor for more than 35 years, developed what he called the "Gabriel Principle" to describe how people sometimes fix things that are not necessarily broken, in order to avoid fixing those things that really are.

While writing this book, I have asked myself a number of times whether trying to distill and address the challenges of an industry as broadly focused as electricity was necessary, and whether my contribution to the solution would fall into the "Gabriel Principle." Obviously, if you are reading these words, I have decided that there are some ideas worth positing and that I might add something to the discussion in attempting to advance the cause of a viable, stable, and powerful electricity business sector for the most critical industry in the world.

In reviewing much of the contemporary and not so contemporary literature on this industry, it became apparent that the focus has generally been given to science and technology, the regulatory environment and legislation, financing and trading or history and economics. Similarly, business literature, both modern and from days past, has primarily focused attention on industries and commercial ventures that live in different worlds from those inhabited by the investor-owned parts of the electricity industry, the public power entities, cooperative organizations, and municipal power suppliers.

As a professional in the industry—and more than a casual observer—it appears that many of the challenges, quandaries, and stumbles that occurred during the past 10 years have stemmed from a host of factors of the sector's own making. Some of these factors are rooted in the basic corporate strategies used by energy companies as they attempted to transition to a competitive environment. The collective health, stability, and success of these very companies is vital, as they directly affect the health, stability, and success of the industry and nation as a whole.

The central thesis to this book has two guiding themes. First, certain business principles and practices need to be adopted for the electric sector to survive and prosper. This is one of the only businesses that has consistently failed at re-invention—that is, failed to change both its structure and its underlying technology infrastructure—and yet has survived relatively unscathed in a rapidly expanding technological environment. It is perhaps a substantial testament to electricity's early pioneers, Edison, Westinghouse, and Insull, that the industry is essentially using the same basic technology 100 years after its initial development. While there is significant talk about a new vision, a smart grid and money being expended in developing new generation technologies, the business is still guided by caution and a refusal to take risks—all at a time when the future of the business and strength of the nation grows increasingly dependent on electricity.

The second critical theme is that there are five fundamental megatrends that have shaped and will continue to shape the future of the business. These megatrends or "destinies" show the way for companies to follow in the pursuit of the future. Failing to recognize and take advantage of the megatrends will result in unsuccessful business structures and a delay in the inevitable.

When I began this project four years ago the industry was slowly coming back from the post-Enron haze, we had just suffered the largest blackout in U.S. history, an energy policy act was still a dream, the industry was battling over FERC's centralization via transmission organizations, investment in infrastructure was at its lowest point since the Great Depression and natural gas prices were in the $4 range. In 2007 energy trading is coming back slowly, we are 18 months into Energy Policy Act with little teeth, the ISO/RTO movement has slowed, investment is increasing somewhat (primarily in advanced metering and transmission); but, we are facing significant generation capacity shortages in the next 3-7 years, and gas is north of $6.

Visions for a Sustainable Energy Future focuses on two essential elements of the electricity business: the basic regulatory/competitive structure, and corporate investment strategy. It then takes the megatrends that will impact the industry for many years to come and outlines how companies can take advantage of those opportunities. It asserts that the industry needs a framework that strikes a balance between competition and regulation. A framework for competition will create healthy markets and profitable companies, while a framework for some level of continued regulation will create incentives for companies to invest in the technology needed to ensure the reliable delivery of electricity. This is an accountability jointly shared.

Looking at the past as a prologue to the future, this book explores how companies in the electricity sector can use their greatest strength—technology—to define a new future for themselves and the constituencies they serve. The electricity business, like every other, needs a solid base of commodity operations. It also needs to take some risk in reinvention and expansion before coming full circle in using that invention as a new foundation from which to build toward the future. Once regulators and energy companies can create a regulatory framework and corporate business model that feeds this healthy balance, encourages a constant reinvention, and develops a reward structure for rate of return (or reduced operating costs for public power), the industry will be on the road to a solid and sustainable future.

The book outlines the five megatrends, describes their primary issues, and provides examples of transformational technology opportunities that companies can engage in through a framework that supports a balanced competitive regulatory scheme, driving business growth. The book goes on to describe the "technology destination" created by the integrated application of all these technologies—a vision of a transformed electricity sector that supports the diverse needs of the 21st century.

Electricity remains an amazing business, populated by a broad spectrum of dedicated people, having a fundamental and decisive role as society's engine for prosperity, and offering a tremendous opportunity for the future. As Kurt Yeager, the former CEO and president of EPRI, was fond of saying, *"It is the answer, not the problem."* Electricity is the answer to an array of global issues, from poverty and pollution to improved quality and vitality in life. This precious, though under-appreciated commodity needs to be supported in a robust manner.

The August 14, 2003, northeast power outage occurred as I started to write this book, underscoring not only the vulnerability of the power delivery system, but also the impacts of all of the under- and mis-investment, the misdirected attempts at developing an effective regulatory framework, and the inability of companies to refine successful corporate business strategies that engage the new business environment. In 2007 the industry is anxiously awaiting new carbon regulations as it deals with significant run-ups in new generation costs—generation that must be built to keep the lights on.

It is important to recognize that the industry involves not just the investor-owned community, but a panoply of players. During the recent outage, municipals, co-ops, and public power entities were equally impacted. And critically, all of the players face the same needs and operational challenges. They all need technologies, solutions to environmental requirements, the transportation of electrons, strong trading partners, Wall Street's financial backing, and support from suppliers.

Rarely in history have we had the chance to stand at such a turning point and recognize it as such. Lyndon B. Johnson never saw the signing of the Tonkin Gulf Resolution as the beginning of a huge change in U.S. society, and John F. Kennedy didn't realize that the Bay of Pigs would change the face of foreign policy. We have the opportunity now of looking forward at the tremendous possibilities for transformation of the electricity enterprise and the huge positive impact this can have on society.

On behalf of my father, and my lifelong effort to avoid an act of the "Gabriel Principle," I am confident this book can help companies take advantage of this unique reflection point in time and do their part in contributing to a dynamic, successful electricity business that drives progress and prosperity throughout the 21st century.

Mark A. Gabriel

Chapter 1

Megatrends and the Energy Horizon

The energy enterprise, from electricity and gas to oil and its potential replacements, is being more radically altered than any time in the last 100 years. Not since the initial discovery and expansion of energy systems in the late 1800s has such a dramatic change taken place—and not since that time have there been as many opportunities and challenges.

By identifying the key elements of this transition and examining the experiences and successes of stakeholders, energy companies can take advantage of the market forces driving change. The goal is to find the overarching, immutable megatrends, analyze current business capabilities in light of those megatrends and build a robust set of practices in support of the changes that energy companies can combine in custom fashion to create successful business strategies.

The concept of a megatrend is simple: it is occurring regardless of efforts to change its outcome. No amount of "will" or "desire," personal, corporate, or governmental can prevent it from happening. Megatrends can rarely even be affected by those forces. They can be nudged in certain directions, but cannot be stopped or altered in any major way and will have an impact on the nature of the industry for at least a decade.

Market Trend vs. Megatrend

It is easy to confuse market trends with megatrends, often to the demise of a company. The dilemma is that those who follow a market trend are always behind. The opportunity lies in foreseeing a megatrend, building strategic direction around it, and giving rise to new sources of revenue, improved ways of doing business, and company longevity.

In the commercial world, there are numerous examples of companies that make money by following market trends, being second to the market with a product or service. These firms recognize the nature of their business in terms of duration and limits to market share. There are also many examples of companies that established a market trend, only to follow it too long to their demise.

In the electricity industry, however, where investments are huge, the product is of a critical nature and the time lines are long, following a market trend can yield particularly disastrous results. The market trend of international expansion is such an instance as dozens of IOUs plunged into ownership of overseas operations during the 1990s only to find themselves operating in unfamiliar countries, with regulatory schemes and governments not well understood and which resulted in financial losses totaling billions of dollars.

It is not only the utilities which follow market trends but often the regulatory and financial communities as well. Deregulation is perhaps the clearest example of a market trend—not a megatrend—because it can be and was changed. This market trend plunged parts of the country into great chaos. The megatrend of customer engagement was forgotten by most companies in the market trend push to customer *choice*. There was a mistaken belief that what customers wanted was a choice of suppliers, when, in fact, customers want a choice in levels of service in order to affect their bills or special power needs. Customers have shown repeatedly that they want to have a role in their energy consumption, but care little, if at all, about the choice of their supplier. As with other segments, such as telephony, internet service or cable TV, customers want the option to select the types of features and services and generally do not care about the specific provider.

Utilities have often been cited as "fast followers" and classic "second adopters" in a Geoffrey Moore sense. This is due in part to the regulatory process and the valid concern about system reliability. It has also meant the pace of change has been slow. It will be those companies that recognize the megatrends of the day and take active advantage of the ensuing opportunities that will survive and prosper.

The Quandary of Scenario Development in Strategic Planning

Planning around these megatrends is significantly different from scenario development. Almost by definition, scenario planning is wrong. Scenarios that are developed are never realized in their entirety. They can be useful from a discussion perspective but tend to the extreme and therefore of limited use. The resultant modeling in scenario development is useful in terms of examining possible outcomes, but creates an environment of looking from today's perspective forward, rather than understanding tomorrow and looking backwards to see what it will take to achieve the desired goals. It is all too easy to miss major changes and opportunities since scenarios are based on assumptions from past experience.

There are numerous instances in the energy business where the industry has fallen into the scenario trap. Whether it was the predictions in the 1970s of oil being depleted by the year 2000, to nuclear power too cheap to meter, to natural gas price stability at $2-3 a MMBtu, the scenarios created expected outcomes based on the past, not on future vision. Traditional business case stories abound of the railroads failing to see the threat represented by the onset of automobiles, buses and trucks; ice merchants making significant improvements in the storage and transportation of ice only to be wiped out by refrigeration, and typewriter/calculator manufacturers sent packing by the emergence of desktop computers. All of these examples demonstrate a strategic decision to extrapolate the future from their existing positions and base forward-thinking scenarios on what they knew and expected to happen. In each case they failed to see the megatrends of their day: personalized transportation, inexpensive electricity, and the transistor.

As Henry Ford was fond of saying, "If I had asked my customers what they wanted, they would have asked for a faster horse." Ford did not extrapolate from the horse, he envisioned transportation in a new way.

If the industry plans its scenarios around carbon taxes (now estimated at $7-$37 a ton by 2015) it may miss the opportunity in carbon injection in saline aquifers; if it plans on scenarios relying on natural gas as its main fuel source it may miss the impact the intelligent infrastructure will have on consumption and the demands of the customers; if it plans on a consistent regulatory compact it may soon find the business model assumptions were wrong. Therefore, rather than scenarios, viewing megatrends is at the heart of building a sustainable energy business for the future.

A Current Crossroads: The Electricity Sector Today

The development of the electric power system, often described as the most complex machine in existence, ranks among the most significant engineering feats in human history—one that has transformed society and the way people work and live. Electricity has revolutionized communications, drives industry and business, and is inextricably linked to the quality of life around the globe. Moreover, electricity has become essential to today's microprocessor-based economy and to public health and safety. In short, electricity is our most critical energy supply system, driving the nation's engine of progress and prosperity. Unfortunately, this energy source, having become such a ubiquitous and reliable part of everyday life to each and every American, has relegated itself to the background of consumer consciousness.

This first chapter outlines some of the key challenges facing the electricity industry, and examines how the history of the industry helped create those challenges, focusing largely on the actions of Samuel Insull, a leader whose innovative business strategies and intensive technology investments created the electricity industry as we now know it. It then explores conflicts in the marketplace that stem from the divergent goals of key industry stakeholders. Finally, it briefly presents a corporate business model that can help the industry cope with its challenges and conflicts and create a context for its future growth and prosperity—issues that will be explored in greater detail in subsequent chapters.

The Challenges

After more than 75 years of regulation, the electricity industry began experimenting with various forms of deregulation in the 1990s. Regulators, energy companies, and customers alike have faced significant difficulties throughout this transformation. Today, the power system and the electricity industry stand at a critical juncture, facing an array of challenges that place the industry in an unstable and ambiguous position. Electric utilities must cope with not only the challenges of competition, but pressures from Wall Street, consumers, government policy, and an unclear economic outlook.

Further, the regulatory scenario governing utility business operations is built on a set of inconsistent and conflicting rules. As deregulation unfolded during the late 1990s and early 2000s, complications were many and the stated objectives of competitive restructuring—lower prices and more choices for consumers—were lost. Certainly, the incentive to serve customers at the lowest possible cost has been overshadowed by near-term financial concerns and diminishing economies-of-scale. Complicating matters, the future of deregulation is fraught with uncertainty.

Electricity companies are also subject to a maze of additional regulatory mandates designed to ensure environmental protection. These rules are complex, changeable, and based on evolving and often conflicting science. Companies are forced to make strategic environmental decisions involving significant investments, while national and state environmental policies and even core scientific issues are unclear and unresolved.

Imposing technology challenges are also at hand. The power system requires critical upgrades, yet the current regulatory/competitive environment has vastly diminished the financial incentive for investing in such upgrades—let alone for the development of advanced technologies necessary to build a power infrastructure that will meet 21st century demands. As a result, capital expenditures by U.S. electricity providers were only about 12% of revenues during the 1990s, less than one-half of historic minimum levels and even below the lowest levels reached during the Depression (Figure 1-1).

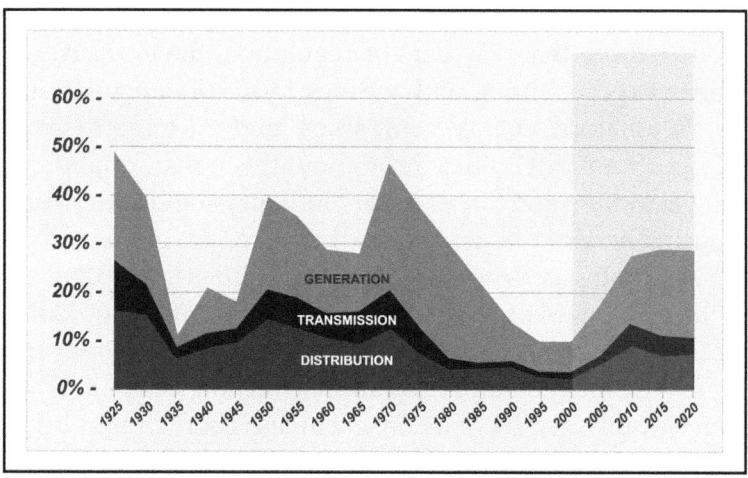

Figure 1-1. Capital expenditures by U.S. electricity providers were only about 12% of revenues during the 1990s, less than one-half of historic minimum levels and even below the level reached only briefly during the Depression. *Source: EPRI Electricity Sector Framework for the Future, Volume I, August, 2003*

Energy Business Capital Expenditures As a Percent of Revenue

According to Utilipoint's analysis of FERC Form-1 filings, R&D investment per utility dropped from about $9 million in 1995 to $3.4 million in 2000 while venture capital investment in the industry inched up until 2005. Since then there has been a huge surge in investment in what is known as "clean tech" arena with more than $18 billion flowing in search of new generation and fuel sources. However, most of these dollars are chasing a small subset of the marketplace in alternative energy and areas such as ethanol, ignoring the traditional thermal value chain as well as technologies to support advanced transmission and distribution. Some money has gone into advanced control systems with recent payoffs in investments.

This historically low level of investment reflects the uncertainty energy companies are now confronting. Many find it difficult to make even short-term strategic decisions, let alone engage in long-term planning. Instead, they find themselves "treading water." In the meantime, consumers and industry continue to demand more and more from an aging, obsolescent power infrastructure.

Mr. Monopoly and the Roots of Regulation

A fundamental underlying reason for the electricity industry's current challenges is the public perception of electricity—an invisible ubiquitous force that will always be available, inexpensive, of high quality, and reliable. To understand the evolution of this perception, we can look at the beginnings of the electricity enterprise and one of the most prominent business figures of the 20th century, Samuel Insull.

At the dawn of electrification in the late 1800s, Samuel Insull was one of Thomas Edison's top executives at General Electric. Envisioning the future of electric control stations, Insull took the new position of president of the Chicago Edison Company, where he became a pioneer of new technology and new business concepts. Thanks to substantial investment in new technology, Chicago Edison had 80,000 homes wired for power by 1912. The company promoted the "all-electric kitchen," electric household appliances, and electric public transportation. In the late 1920s, Insull invested in rural electrification, bringing power to remote areas and setting the model for today's power grid. Through this heavy investment in transmission, Insull not only brought power to the people, but was also able to increase the generating efficiencies of his larger units, slash rates, reduce operating costs, and increase load factors—enhancing profitability in the process (Figure 1-2).

By 1930, Insull had assets in excess of $2 billion and produced one-tenth of the nation's electricity, offering service to 5,000 communities in 32 states. Insull's economic business model strove to

Figure 1-2. Samuel Insull's investment in technology played a key role in developing the electricity enterprise as we know it today. Even with his insistence that the electricity industry be regulated, his success led him to be immortalized as Mr. Moneybags in the classic American game, Monopoly. *Source: Loyola University of Chicago Archives: Samuel Insull Papers*

extend electric service to as many people as possible at the lowest possible cost. This he termed "massing production," predating Henry Ford's "mass production" of the Model T.

Insull's investment in technology played a key role in developing the electricity enterprise. He was a pioneer of new business concepts in energy and transportation and was responsible for many advanced technology developments and financial business practices fundamental to the electricity industry's rise. Among them:

- Mass production using large-scale generation stations,

- AC to DC conversion to improve transmission capacity,

- Expansion of the power grid for rural electrification,

- Using the Wright Demand Meter to discount non-peak rates and decrease residential rates, and

- Accounting practices.

In expanding rural electrification and using technology to bring electricity to the people, Insull not only helped create the modern electricity industry, but also started the nation toward the perception of electricity not a privilege, but an entitlement. Even Roosevelt's New Deal activism (Figure 1-3) upheld that the large federal hydropower projects "…will be forever a national yardstick to prevent extortion against the public and to encourage the wider use of that servant of the people—electric power." Many companies throughout the development of the enterprise reinforced this perception through even their names, which often included the words "Public Service."

Insull believed that a regulated monopoly was the only way to run an energy company fairly and to provide a source of guaran-

Figure 1-3. Although Insull maintained a good relationship with Washington and President-elect Roosevelt, his electric utility holding company business model was eventually denounced as the "Insull Monstrosity," leading to far-reaching legislation. *Source: Chicago Historical Society (Negative #DN-0076539)*

tees to ensure investor confidence. As early as 1898, in his presidential address before the National Electric Light Association, he lobbied for regulation by state commissions. He proposed that states be given full power to fix rates and standards of service, and "...to alter the conditions of franchises so that if a company failed to offer satisfactory service, the municipality it served would have the right to acquire its plant at cost, less depreciation..."

When the stock market crashed in 1929, Insull's securities dropped tremendously due to excess debt. Although his business was healthy, his New York creditors claimed his empire was "on shaky ground." On September 21, 1932, Democratic presidential candidate Franklin D. Roosevelt delivered a speech as part of his New Deal activism. Presenting his power policies, Roosevelt denounced the "Insull monstrosity" and implied that other public utilities operated in the same way as Chicago Edison (Figure 1-3). He was referring to the large holding company structures that were growing in popularity and controlling more and more of the utility and other industries. At that time, only eight utility holding companies controlled almost three-quarters of the investor-owned utility (IOU) business.

Insull lost everything to the federal banks. He was forced to resign from 60 corporations and, facing indictment for mail fraud, bankruptcy violations, and embezzlement, he left the country. In retrospect, Insull was probably set up as a scapegoat for the financial woes of the country. After several trials and losing his entire fortune, Insull was acquitted on all charges and died penniless in France.

Insull's alleged wrongful business practices eventually led to some of the most far-reaching legislation in the history of the U.S. congress including:

- Public Utilities Holding Company Act (PUHCA) of 1938,
- Acts creating the Tennessee Valley Authority,
- Securities Exchange Commission, and
- Rural Electrification Administration.

Though largely forgotten, Insull forever changed the face of the electricity enterprise. Drawing on his legacy of innovation—rather than his legacy of physical and regulatory structures—could help the industry find and exit from many of the challenges it now faces.

Inherent Conflicts in the Energy Business

Today, the pushes and pulls of the marketplace have created an environment of inherent conflicts, as clarified by the perspectives of the four major stakeholders in the electricity enterprise: investors, consumers, generators, and power marketers (Figure 1-4). What does each of these stakeholders want?

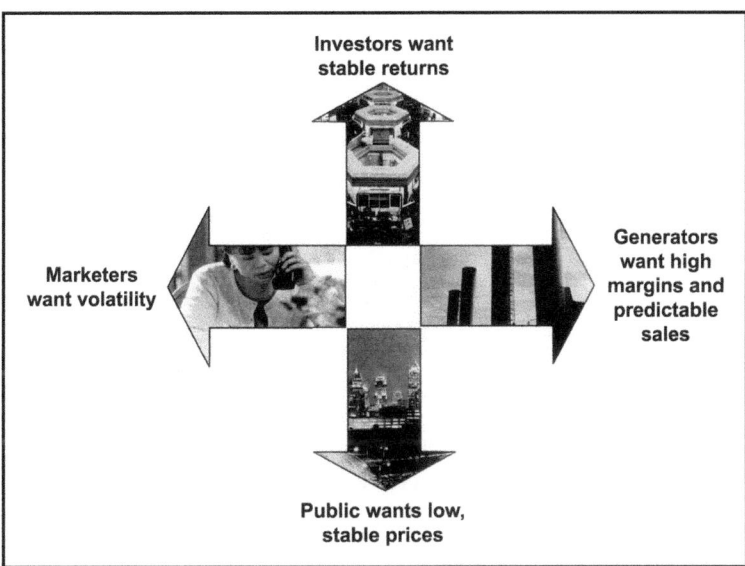

Figure 1-4. The pushes and pulls of the energy marketplace have created inherent conflicts between investors, generators, power marketers and customers, immobilizing progress of the industry to move out of the current dilemma.

Inherent Conflicts: Pushes and Pulls
Investors

Investors want stable and predictable returns and predictability coupled with the potential for upside stock appreciation. This aspiration has been supported by the 2003 Tax Reform Act, which reduced the tax on dividends. This change is particularly important for the utility sector, as it is classically the largest payer of dividends in the stock market. On the other hand, from an investor standpoint, what looks like a good return today—when people are starving for returns—pales in an era of double digit returns, as has occurred with high tech stocks.

Consumers

Consumers want low, stable prices. The electricity industry has succeeded so well in becoming a ubiquitous background force while keeping prices down over the years that customers have forgotten the value of electricity service. The same consumers who are willing to shell out $150 dollars each month for the convenience of a cell phone cringe at the thought of increasing their electricity bill by $10 per month to increase the comfort of their home with more efficient, load-controlled air conditioning. The public's reluctance to recognize the value of electricity—and therefore pay for that value—stems largely from the widespread perception of the ready availability of low-cost electricity as an inalienable right.

Generators

Generators want high margins and predictable sales to increase profits and satisfy their shareholders. However, there is limited profitability in the electricity business, partly because the regulatory compact sought to ameliorate the ups and downs in the market, keeping prices stable for customers. Even non-regulated power providers, who can capture higher returns than a regulated utility in a high-demand market, are at a higher risk in a low-demand situation. This limits excessive profits as the number of days a year that demand is high is limited.

Generators do want some volatility, because creating shortages—either artificially (via trading) or through practical reliability (through constrained systems) makes money for generators, regardless of their regulatory structure. This is not only true for the IOUs, but for public systems as well. Municipal systems contribute significantly to their communities' budgets; rural co-ops continue to make money for their "owners" (the publics they serve), and the federal providers bolster their operations in building, managing, and growing their areas of control. Too much volatility, however, makes it hard to meet the investor's demand for stable returns. During California's 1999/2000 winter "shortage," the Los Angeles Department of Water & Power sold significant quantities of generation, at a sizeable profit into the marketplace.

Power Marketers

Volatility is a power marketer's friend. The power marketer sector, which has been almost entirely wiped out in the return to asset-based trading, truly benefits from market volatility, up or down.

How the Conflicts Play Out

A quick comparison of the aspirations of the various stakeholders shows direct conflicts:

- Investors want high returns that generating companies have difficulty supplying, since generating electricity is not a highly profitable or exciting business. Disappointed, investors are placing their funds in other, more profitable businesses.

- Generators are in conflict with consumers because they want rates to go up and consumers want the rates to go down—a perspective supported by the regulatory framework for decades. This price constraint has created a precarious situation for the industry, in the face of rising operating costs and an aging infrastructure.

- The power marketers' quest for volatility puts them in conflict with generators, investors, and consumers. Volatility increases the difficulty of satisfying investors happy with consistently high returns, and drives up energy prices, anathema to consumers.

Relieving the Conflict

The conflicting pushes and pulls of these four parties have immobilized the progress of moving out of the current dilemma. If IOUs and cooperatives try to maintain the old model of rate basing all activities and squeezing margins to pay a return to investors, they will continue to prolong a paradigm that is no longer functional. Sweating the assets is a short-term solution to a broader problem. Virtually all stakeholders in the industry agree that continuing a business as usual approach is unsustainable, unprofitable, and unimaginative.

Further, the regulatory price constraints limit the ability of system operators to invest in the in aging infrastructure. There is growing concern that, unless the industry's physical assets are rebuilt, they risk either collapse or rejection as a relic of the 21st century. It seems unreasonable and undesirable that distributed energy resources (DER) could supplant the nation's immense T&D infrastructure. We must keep in mind that past lessons are often forgotten—lessons such as how the automobile, a distributed resource, displaced the railroads.

However, any solution to the industry's challenges will not ultimately provide for the long-term health and balance of electricity companies or the industry as a whole unless it encourages companies to adopt healthy corporate strategies. Most industries thrive on a healthy balance between three cyclic activities:

Foremost, companies operate a relatively safe core commodity and service business that provides a dependable rate of return. Revenue from this commodity and service business feeds new technology development. In turn, new technology becomes the basis for new, more risky business opportunities that build on the company's core business—essentially allowing the company to reinvent itself with new commodity and service offerings. Profits from these new offerings fund further technology development and the reinvention continues.

This book will demonstrate that the collapse of Enron and the challenges facing many electric energy companies are symptoms of an unbalanced corporate strategy focused on market trends. Acknowledgment of current market forces and balanced strategic planning around the current megatrends are the critical drivers to build success for the future. Drawing on experiences of other deregulated industries, the chapters ahead will present a corporate investment strategy model for energy companies and a balanced regulatory model that supports and incents corporate investment. Once energy companies can act upon a business model that feeds this healthy balance, by developing a reward structure for necessary innovation to encourage investment and constant reinvention, the industry will be on the road to a solid and sustainable future.

The electricity industry urgently needs a new vision for a revitalized energy sector that reduces uncertainty and improves confidence through stable and predictable regulation and rates of return, clearly defined accountabilities, and open and transparent markets for all customers.

Companies need to look at new horizons to create value and build business—drawing on the experiences of other industries for their inspiration. To make the shift from business as usual and toward a new vision for the industry, new regulatory and financial models that support such transformation are essential. The chapters ahead will explore possibilities for those new regulatory and financial models.

In looking across the enterprise, there are five megatrends that appear to meet the simple criteria of immutability. These trends are major in scope and in their affect on industry, companies, and society—and they are occurring now. Each one of these trends spins off in a number of directions and has subsets or mini-trends. While there are variations on these trends, the basic themes remain. The question is how a company can best position itself to take advantage of these megatrends and turn them into business opportunity. The five megatrends are as follows:

- Business Model Evolution
- Aging Workforce and the Boomer Effect
- Carbon Constraints and Capacity Challenges

- Intelligent Infrastructure
- Customer Engagement

In examining each of these megatrends, it is important to recognize the current political, financial, social and technical forces that exert an influence on the trends. These are merely reactive factors, not causal.

- Workforce retirement ages may be pushed ahead but the population still ages.

- Increased public concern over climate change will mean stricter regulations, yet electric power requirements will need to be met regardless.

- Regulators may incentivize utilities to build the smart grid (and should) but utilities, oil companies and alike are bridging current infrastructure with communications and computing intelligence regardless, because it is proving to be increasingly more efficient for economic reasons.

- Open retail access may be delayed or expanded but, regardless, consumers are and will increasingly demand differentiation and levels of service.

- The drive to new business models is stimulated both through economic necessity and strategic planning around the megatrends.

Another key element of these megatrends is the coupling of immutability with a long time frame. These are issues that are not quickly or ever "resolved." There may be periods of rapid change or slow movement, but the megatrends occur nevertheless. There is no short-term "end" to a megatrend—just waypoints down the path. According to Jim Nesbitt who pioneered the concept of megatrend analysis, these megatrends last at least 10 years. In building a sustainable energy business, the five megatrends will persist for the foreseeable planning horizons. This means that company planning must constantly continue its evolution and can never be content that a final solution has been found. Successful businesses anticipate and plan towards the megatrend as part of their journey—unsuccessful businesses fail to take advantage of the opportunities and therefore stagnate and are subsumed.

This book looks at these five megatrends affecting the energy enterprise today and suggests many investment opportunities that can propel a company forward when planned and executed through a balanced corporate strategy of investment. When taken enterprise-wide, the result is the successful expansion and continuation of the electricity sector. In the next two chapters, we will explore how these megatrends can be approached by power producers and power providers through a competitive investment model and through a balanced business strategy. In this way, companies can turn what seems like an overwhelming challenge of change into equally powerful opportunity.

Chapter 2

Incenting Innovation: A Competitive Regulatory Framework

Developing a sustainable business framework for the electricity business involves complex issues which are affected by deregulation, market forces and the role of the customer. By looking at a brief history of electricity deregulation, examining how various electricity companies have responded to deregulation and how their response differed in both strategy and success from those approaches taken by companies in other industries we can begin to see a pattern of success through core investment. This comparison solidifies our position that investing in innovations to the core business infrastructure is the only viable strategy. A regulatory-competitive framework that creates incentives for infrastructure innovations (through a regulated return on investment), permits sharing of risks and rewards between regulators and companies for some medium-risk activities, and leaves companies free to pursue high-risk, high reward activities on their own. In addition, because deregulation to date has largely shut out the customer, the role of the customer will need to transform in this new model.

Turning a megatrend into business opportunity is the key in the context of a balanced corporate strategy. Throughout this book, the innovations and solutions provided for each megatrend will fit into a competitive regulatory framework that creates the incentives, market risk and commensurate return on investment that companies will need to embrace each megatrend and use its power for full business transformation.

The Regulatory Dilemma

Regulation served the investor-owned power industry well through the 1960s. Due to enormous investments in the 1920s, 1950s, and 1970s that built the electricity enterprise, utilities were able to use improved technology to exploit economies of scale and provide inexpensive power to customers while receiving high returns. As the 1970s unfolded, however, the U.S. energy business faced new challenges—including those resulting from the oil crisis, higher fuel prices, and inflation—that drove electricity costs and prices upward. Environmental impacts became an added concern as the financial burden of investments in environmental protection increased.

A major tectonic shift occurred in 1978 with the passage of the Public Utility Regulatory Policies Act (PURPA). Intended to encourage more energy-efficient and environmentally friendly energy production, PURPA defined a new class of electricity producer—the qualifying facility (QF). QFs (the forerunners of today's independent power producers) are defined as either non-utility entities that generate energy for their own needs as well as occasional or frequent surplus energy, or entities that happen to generate usable electric energy as a by-product of other activities. Under PURPA, utility companies are obliged to purchase energy from these facilities based on a pricing structure referred to as avoided cost rates.

PURPA was the first crack in the regulated monopoly model and set in motion the forces of power industry deregulation. Soon thereafter, in March, 1979, the well-publicized accident at the Three Mile Island nuclear plant increased the public's negative perceptions of nuclear power, while undermining the credibility of "Big Energy" and the vertical monopoly utility model.

The ground continued to shift in the 1980s, as utility regulators espoused a new concept of financial recovery with the introduction of the "negawatt"—or power that the utility could avoid producing by introducing some kind of conservation or efficiency measure—and a new perspective on demand vs. supply. Resulting rules of demand-side management (DSM) allowed utilities financial recovery for *not* selling electricity through a standard rate of return. In fact, DSM can be an excellent tool to manage loads and provide energy-efficiency incentives if properly applied, and consumers have shown a willingness to participate in DSM, provided they can see a direct connection between energy savings and monetary compensation. Unfortunately, the design of many DSM programs—intended to increase efficiency and reduce consumption—seems to have been compromised by the need to appease regulators. In many cases, these programs actually increased costs, promoted inefficient use of resources, and failed to provide known, measurable, and lasting solutions.

The Advent of Deregulation

The momentum of deregulation in other industries—with the promise of lower prices and more consumer choice—began to pull at the electric power industry. Many industries were deregulated between the late 1970s and the mid-1990s, including airlines, railroads, trucking, financial services, telecommunications, and natural gas transportation. Deregulation took various forms, depending on the nature of each industry and their corresponding regulatory regime. In general, results were good: regulatory reform lowered prices and increased efficiency, productivity, price discrimination, product variation, and overall consumer welfare. In addition, profits appear to have risen in the telecommunication and airline industries immediately after the removal of regulations. And in each case, as new technology opened up the door to consumer choice, service offerings mushroomed, and entrepreneurial activity went into high gear.

Hoping to see similar benefits in the electricity enterprise, as well as relieve mounting stresses, states began to examine utility regulation, with California at the forefront. By 1994, electricity rates in California were among the highest in the nation, double those of surrounding states, and consumer groups were pressuring the governor and legislature to lower electricity prices. Utilities added their weight, looking to deregulation as a way to resolve issues of stranded assets, unlock corporate value for shareholders, and position themselves in a new future. And many industrials were leaving the state, investing in new facilities outside of the state, or lobbying to purchase power from sources other than the state's big three IOUs. In 1996, California's legislature unanimously passed AB 1890, and deregulation was signed into law.

While California went boldly where no one had gone before, the movement toward deregulation in other states was progressing. States dabbled with changes—Texas, for example, deregulated in 2002 and other states continue to experiment with versions of deregulation. To help standardize markets across the country, the Federal Energy Regulatory Commission issued their Notification of Proposed Rulemaking for Standard Market Design (SMD) in 2002.

But California's experiment was not the success the industry was hoping for (Table 2-1). Although thousands of retailers made offers, by December of 2000, less than 2% of the California residential market had switched suppliers, and only 12% of industrial customers had switched. However, the real blow to deregulation came when series of supply shortages in the winter of 2000-2001 sharply increased residential and business energy bills, sending the state into economic and political turmoil.

As the state government tried to protect California energy users and help bail out the now near-bankrupt utilities, the federal government finally capped prices. By 2002, the power marketing sector—widely implicated in California's energy crisis—was in shambles, and by 2003, FERC's SMD was floundering amidst a slow economic recovery and the ongoing specter of utility credit issues. In short, the industry as a whole is still grappling with the complex and confusing issues of deregulation.

In the period between 2003 and 2007, many jurisdictions stepped back from deregulation and others are considering their alternatives as much anticipated savings never materialized. Faced with increasing generation costs, significant financial demands on infrastructure and increasing price pressures have made the thought of further deregulation a political non-starter. The reality of rate caps coming off across the country and the resulting price shocks

Table 2-1. The goals and expectations of electricity restructuring did not materialize as planned and the current reality in the sector is not ideal. Electricity companies have taken a number of tacks in response to this situation.

Vision or Reality:
Expectations and Current Reality of Electricity Deregulation

Vision	Reality
Lower the cost of reliable, safe, clean electric service	Higher cost and lower reliability
Attract capital for infrastructure development	Reduced investment confidence and incentive
Enable greater consumer choice	Loss of accountability to consumers
Greater economic efficiency	Greater financial risk
Level the competitive playing field	Market volatility

Source: EPRI *Electricity Sector Framework for the Future*, Volume I, August, 2003

have focused attention on the promises made in the deal for asset recoveries. In Maryland, the request for a 72% rate increase scuttled a proposed merger for Constellation Energy; in Illinois, Exelon's price rises have created a standoff with regulators, legislators and the public, and in California, residents are paying, and will continue to pay, for uneconomic long-term power contracts that were necessary to bring the state out of its energy crisis.

Public Power and Regulation

While public power—both municipal systems and rural electric cooperatives—tried to stay away from the fray of deregulation, they have been inevitably caught up as the industry moved and changed. Exemption from participation in open markets in many jurisdictions has not exempted public power from the machinations of the market, whether it is transmission congestion, system reliability issues, the availability of power supply or the demands of their customers. During some of the challenging years in the west in the early 2000s, public power was able to sit out the market fluctuations, but was impacted nevertheless as supplies dwindled and prices escalated.

The 2005 Energy Policy Act and recent actions by the Federal Energy Regulatory Commission are forcing more players into the same regulatory arena as their IOU counterparts. Issues such as independent system operator (ISO) participation mean identical regulatory costs and the necessity to act in a similar manner. EPACT requirements on all "utilities" selling more than 500 GWh per year at retail mean the same examinations of advanced metering, time of use rates, interconnection standards, fuel choice and generation mix—and the same potential investments. NERC standards on security and reliability cut across all players in the energy enterprise, regardless of IOU, municipal or cooperative power status.

Though sometimes delayed, a number of markets such as Texas are moving to bring municipal power into the broad picture—even as deregulation fades as a market trend for IOUs. Thus, regulated by states or not, public power is finding itself in many of the same situations as the IOU community. Regulatory mandates on power production and the reduction of carbon and other emissions play across the industry on a non-discriminatory basis. This is also reflected in the financial community's view of power in general—regardless of the source.

Electricity Company Responses to Deregulation: Invest in the New

Throughout 1995-2000, hoping to take advantage of deregulation and fired by visions of sexy Wall Street returns, IOUs began to break out of the traditional mold. Pressures from shareholders and the investment community during this period were particularly intense. Capital in the marketplace was fleeing from stable but boring dividend-based investments (i.e., utility stock) as investors turned to dynamic, high-tech, high-spec non-dividend stocks. To woo back and entice new investors and show double-digit returns, energy companies increasingly turned to investing in strategic ventures outside of their core business—in everything from insurance to fiber optics. Overall, their track record failed to impress.

The primary reason for the disappointing outcomes was that in investing in new ventures, utilities neglected a key strategic ingredient: their core business. In embracing the "new," they lost the faith of their investors, their regulators, and their customers. Some lost everything.

Although misguided, the energy company response was not atypical. As industries deregulate, most market players will coalesce around a new strategic paradigm until a market weakness is discovered and exploited. The rush by non-regulated generation arms of traditional utilities to buy power plants at above market prices, while their parent organizations sold off power plants (often by regulatory mandate), resulted in numerous bankruptcies, power supply challenges, and financial downgrades. The same was true as many utilities opened power trading operations and found themselves upside down in a volatile market. The paradigm then unravels and another new market structure is created. Over the next decade, the electricity business is likely to traverse this cycle several times before a new stable marketplace emerges.

Lessons from Other Industries:
Invest in Core Infrastructure Innovations

How did the airline, telecommunications, and other deregulated industries overcome this cycle and create stable markets? How did their investments support the health of their companies and industries as a whole? They invested in technology and methodologies to reinvent their businesses in the wake of deregulation or simply as a consistent corporate strategy. *Their investments supported their core business.*

The call-waiting model—a phenomenal success for the telephone industry—provides a telling example. Predicated on investment on technology infrastructure, the call-waiting product was developed only after the technology changed. Firms did not invest in the infrastructure in order to sell call waiting; call waiting was an outgrowth of the investment. Likewise, in building the cable infrastructure in the 1970s, the telecommunications industry laid the foundation for not only cable, but broadband internet. Investing in information technology, the airlines developed a new reservation system that created a basis for not only better customer care, but improved load management as well. (These and other strategies will be discussed further in Chapter 3, The Destiny of Business Transformation.)

The electricity business has repeatedly tried to discover and launch new business options stemming from its existing infrastructure. Many energy companies, for instance, created new subsidiaries related to their core business. For example, some created energy services companies (ESCOs), convinced that there was a market for services to help businesses increase their energy efficiency. Others invested in ancillary businesses, including home security, HVAC, and building maintenance operations through energy service providers (ESPs), adopting such business practices as performance contracting. Even companies in the public domain—co-ops, municipals, and public power—sought to take advantage of alternative revenue sources. The National Rural Electricity Cooperative Association (NRECA), for example, created Touchstone, a business to help co-ops market their services.

For all these companies, the hope was that selling energy-related products was a path to profitability—their "call waiting." However, unlike call-waiting, not all of these paths are based on new technology infrastructure investments. And unlike call-waiting, not all of these paths have been successful. Even the successful companies have failed to equal or surpass the regulated rates of return

The irony today is that the utilities that stuck with a steady rate of return are surviving and reaping the rewards of their patience in the short term. Those that had forsaken rates of return and strayed from their core business are in wide ranging attempts near or in bankruptcy and looking back at the good old days with their stodgy but steady commodity business.

Finding an Appropriate Investment Model While Satisfying the "Street"

One has to question whether the electric utility business, built on a cost-plus model, will ever be suited to a margin business. There is very little margin in providing the "value-added" services offered by the new ancillary businesses, and by its very nature, electricity service has not been and for the most part will never be an "exciting" investment opportunity. In fact, for a consuming public unwilling to deal with price fluctuations and a regulatory model that focuses solely in fixing prices as low as possible, the idea of truly allowing the market to determine profitability and opportunity is probably not feasible.

This isn't to say that there aren't enormous opportunities in the electricity business, but rather that such opportunities need to be matched to the profile of the industry and the consumer base it serves. The various models that have grown up around the electricity business can allow for solid rates of return, balanced investment strategies, and significant predictable (and therefore bankable) business strategies regardless of the model chosen—public, private, or government.

A significant financial conundrum exists for the industry in the need to meet or exceed Wall Street expectations and the reality of the current business models under which the sector labors. The twist is in the need to both "sell" the street (and hence investors) on the differing value that energy companies provide (through the combined benefit of dividends and stock appreciation) and to convince the markets that comparisons to other industries should be done on this basis, not on stock price alone.

Through the "irrational exuberance" days of the late 1990s, many investor-owned utilities as well as public power tried new avenues to increase their attractiveness to the marketplace. Public power looked to make itself appear even more solid through related ventures (Direct TV got its start in the co-op market) to drive borrowing costs even lower. Many of the Wall Street pundits challenged IOUs to spin out and unlock value to the market creating a frenzy of activity in what may have been an illogical move.

In order to free capital as well as meet the increased pressure not to raise rates, some in the IOU community changed their dividend payout ratio. While logical at the time, this led some yield investors to drop or reduce their utility portfolios. For a long while, the anticipated run-up in IOU stock failed to materialize for most companies, and thus, the triple whammy of reduced dividends, limited stock price rises (in some cases significant decline) and the loss of some key business adjacencies (Figure 2-1).

The financial good news for some utility stocks was the general market's movement upward as well as the bond market's direction. The rebound of many stocks has been fairly dramatic as the misadventures have been sorted out such as TXU's now abandoned foray overseas and PG&E's rebound from the California energy crisis. But all is not rosy as some of the market participants have recently seen such as Aquila Energy's being forced to sell all of its assets after disastrous plays in the trading a power markets.

In fairness, comparing utility stocks performance to those of other industries is not apples-to-apples because of the dividend factor. Yet, in the search for capital, the markets are often an unkind partner, and the investing public's sophistication in making such distinctions can be limited. As the population ages and demands more from dividend paying entities, it will be a challenge to meet that expectation, invest in new technology and infrastructure, and support growing needs of sophisticated market operations.

The problems of other deregulated industries demonstrate that the dismantling of comprehensive regulated industries should not mean that government is totally hands-off. In fact, many of the failures over the last 15 years have been failures on the part of government to successfully fulfill their responsibilities. Similarly, the electricity business will probably always require some level of regulation, to ensure both that customers continue to have access to electricity as a public good and that electricity companies cooperate on issues essential to the smooth supply of power.

Stock Appreciation of Electricity IOUs, S&P 500 and DJIA Total Return
January 2002-December 2006
(Reflects Reinvested Dividends)

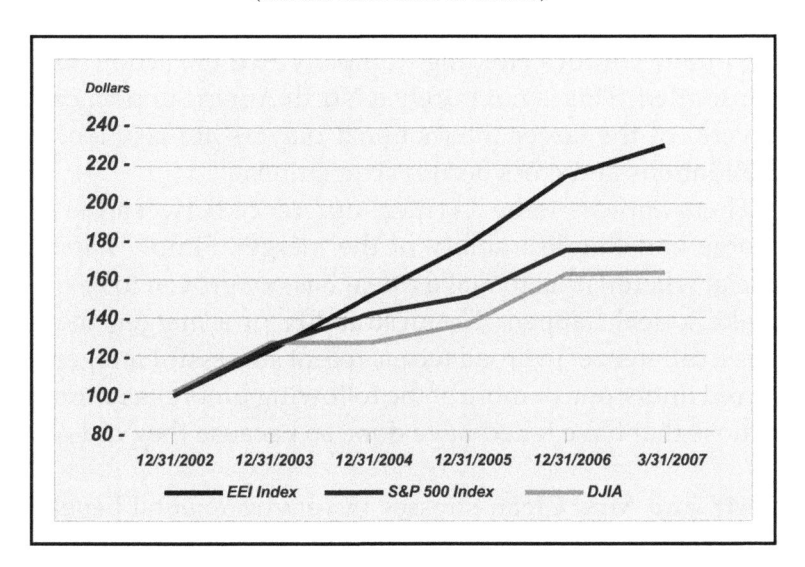

Figure 2-1. While IOUs were challenged in attracting investors during the boom years of the 1990s, the higher risk strategies they adopted were not necessarily suitable business models for such a critical service as electricity—thus the substantial downturn of stock values through 2002. Additionally, comparing these utilities to other, more volatile companies such as Wal-Mart or Intel is not an apples-to-apples comparison as there is a strong case for the stable utility investment. Most consumers would not be happy with even an unregulated IPP which provides power to their local Retail provider engaging in high risk activities that may result in higher energy prices. *Source: EEI 2006 Financial Review*

Mergers and Acquisitions

Investor-owned utility M&A activity in recent years also provides an interesting look at evolving business models and strategy that is often rooted in a market trend and not a megatrend. The recent failure of a number of proposed mergers (Constellation/

FPL, ConEd/Northeast Utilities, as well as the successes in others such as Duke/Cinergy and Kansas City Power and Light/Aquila) shows the continuing evolution of the business. The "urge to merge" was a market trend in the 1990s and early 2000s and, despite predictions of massive consolidation following the repeal of the Public Utilities Holding Company Act, the action has been rather limited. This is not purely a North American phenomenon as several of the larger international players are also looking at consolidations and cross border acquisitions.

There appears to be a critical difference between the ability to merge and the desirability of the merger. Simply having the financial wherewithal to make a deal has not proven to be enough to make a deal happen. The desirability of a merger, therefore, makes a difference. In broad terms, recent successful mergers have occurred under one or more of the following umbrella rationales—and those that have failed have done so because they did not.

- **Mr. and Mrs. Clean** stresses the environmental benefit of a merger. A company with a higher polluting generation mix desires a partner whose generating assets are the cleanest in the nation. The mixture of a coal heavy portfolio, for example, with a clean one can improve a company's green image and help with regulatory compliance, making a merger attractive.

- **Been There, Done That** is financially and regulatory driven. In this approach, a company's experience with a major financial calamity such as a bankruptcy, retention of employees, dealing with Wall Street and managing regulatory relations through troubled times and the significant stock rebound offers a company expertise the merging company could not find easily.

- **Daddy Warbucks** represents the financial strength that a company can bring to a target acquisition. The combination of many IOUs rejuvenated financial position coupled with a significant cash reserve and positive forward outlook make partnering a logical choice to offset weaknesses in a target company's own financial position or to provide capital needed for investment and/or expansion.

- **Bring Me Your Huddled Masses** stresses the value of combining numbers and types of customers into a larger entity to develop economies of scale. This can also be reflected in the provision of needed expertise and savvy in customer marketing and communications.

- **Balancing Act** takes into account the benefit a merger could have on system diversity. A system's strength of encompassing a large geographical area with a variety of weather conditions allows the merger partner a greater diversity to offset the challenges of a smaller, regional system. The generation mix also brings both systems in better balance to more effectively address regulatory concerns and fuel volatility.

- **This *is* the Way to Run a Railroad** recognizes the technical strength of the organization and allows the acquired company to operate in a more efficient manner. In this case in particular, there are mergers which bring benefits back to the acquirer from the target company.

Mergers and acquisitions on their own have not necessarily been a "silver bullet" either in terms of operating efficiency or in stock prices. The mid-1990s philosophy that in order to be competitive, a utility needed to have $3-5 billion in revenue has not proven to be the case. However, mergers and acquisitions that fit the key criteria of mutual desirability will undoubtedly continue to impact the market.

Encouraging Innovation

If, as we've argued, the utility industry requires some level of regulation, the model for that regulation must allow utilities to meet today's business and market realities. Specifically, as noted, the regulated model needs to encourage innovation. The pressures of a competitive (and challenged) marketplace, an unresolved regulatory and political climate, environmental hurdles, energy security, and an aging infrastructure that strains to deliver even the most basic service to a digitally advanced society make it plainly evident that technological solutions are essential to the recovery of the electricity business.

In fact, in most industries, it has become risky *not* to innovate. Every industry is experiencing change at an increasingly rapid rate. The importance of change was highlighted by ex-CEO of General Electric, Jack Welch, who said, "I am convinced that if the rate of change inside an institution is less than the rate of change outside, the end is in sight." Many industries, the utility industry included, have experienced some hard lessons in change.

Even in a regulated environment, there is a demand for company balance. The need to ameliorate upward cost pressure, and increase and maintain high levels of customer service, exists in any form of business. For instance, the provision of power quality services under regulation allows for a more balanced business approach, providing the customer pays for it. These and other services can offset the need to raise rates while compensating the regulated business. For years, such services were a below-the-line activity that provided revenue streams that were excluded from the rate-making process, yet indirectly helped the financial health of the company—a symbiotic relationship in balance.

Similarly, other industries—the airlines, telecom and cable—were able to add related services that customers wanted, were willing to pay for, and eventually expected as a normal customer option. As noted above, the fundamental reason these industries succeeded in new technological services where the electricity business has not is that these industries invested in new infrastructure first and then used that infrastructure to launch new related business options to their customers.

Finding the Right (and Profitable) Customer Compact

Because the heart of any business operation is the customer, it's regrettable that in the process of electricity deregulation, the customer has been lost. Although deregulation was heralded as a customer benefit intended to decrease prices and offer choice, in reality, the customer has not been engaged in the discussion. In fact, the people with the most to gain—customers—have been removed from the picture, while those with the least to gain—regulators and system operators—are in control. When it comes down to it, electricity consumers are the ultimate financiers of the electricity system, and any business model that neglects to place the customer at its core is doomed.

The failure by electricity companies to renew the regulatory-customer compact could be a sleeping giant for many companies. It's true that many consumers were not ready to switch providers in the onset of electricity choice. However, the industry's inability to reestablish the level of customer service created in the 1960s and 1970s may eventually create opportunities for new providers as customer dissatisfaction with service and price rises.

The electricity sector has assumed that consumers want an active role in energy supplier selection, and will shop for energy providers and related services. This is not necessarily the case. Most customers want first to keep the lights on, second to pay low prices, and third to experience no hassle. It stands to reason that customers may prefer to engage in market activities—such as responding to time-of-use pricing and energy management—that help them achieve these goals—rather than choosing a provider, which could be perceived as a hassle. For years, the industry has done well in engaging consumers in various buying decisions—from off-peak water heating programs to air conditioner control schemes and load shedding—that offer customers both choice and rewards. Consumers appear willing to participate in service choices in return for a financial gain, cost reduction, or enhanced comfort.

For most residential and small commercial customers, there is no connection between the quality of electricity they receive and the price they pay. Yet segments of the market have enormous needs for enhanced quality. For example, hospitals and emergency responders and industries such as banking and semiconductor manufacturing, have highly sensitive loads and huge needs for high quality power and availability. Industries in the United States lose some $120 billion each year—1% of the national GDP—due to power outages and irregularities (See Chapter 5, Figure 5-1). Because the existing, aging infrastructure cannot provide the necessary reliability, in many cases these industries have invested in their own advanced technology to mitigate power disturbances and increase productivity. This provides a tremendous opportunity for the electricity business to provide high-reliability, available power and other services valuable to these companies.

The Challenge of Meaningful Customer Engagement
The customer today is fundamentally disenfranchised from the process of electricity. Until consumers become full partners in the electricity marketplace, the need to protect them will continue to foster the notion of entitlement and distort marketplace dynamics, regulation, and the efficient use of energy and capital resources. Regulators need to move from a role of protecting the customer to a role of protecting markets. The challenge for the electric industry is to engage the customer in an open, two-way partnership, fully understanding customer needs and wants, offering a meaningful portfolio of ancillary services—the "call waiting," *per se*—and, in turn, gaining their recognition of the intrinsic value of electricity.

Through infrastructure investment, the electricity sector will experience a paradigm shift that will increase the functionality and value of electricity via consumer benefits that far outweigh the cost of that investment. It will be imperative that customers begin to realize the tremendous benefits of electricity and fuel their willingness to support the necessary infrastructure investment strategy needed to move this industry forward.

Turning a Megatrend into Business Opportunity

The energy business continues to recover from its experiences in deregulation, misguided investment, a hesitation to innovate and invest in core infrastructure, and the challenges of the regulatory-customer compact. This struggle has resulted in a host of symptoms in the industry including a looming lack of qualified workers, a growing conflict of generation capacity and climate change, severe grid constraints that have led to some of the most costly power outages in U.S. history and a disenfranchised customer with increasingly sophisticated needs and demands for choice. As a result, the industry is faced with the need for reinvention and a significant level of business evolution.

The competition for customers and the need to maintain system integrity place all players in the market—IOUs, municipals and co-ops alike—in the same challenging position with the same opportunities. Across the nation municipal entities are bringing fiber rings and advanced communications to their communities; rural electric co-ops continue to be on the forefront of new technology introductions (as they did with Direct TV) and IOUs continually look for opportunities.

Ultimately, all companies will have to undergo transformation; however, they have a choice on how to approach the current and evolving landscape and a choice as to whether the changes are foisted upon them or whether they will be proactive in the market. For an industry unused to taking risks, there is a certain peril in the unknown and untested—but there is similarly peril in waiting for the market to push change.

Historically, many power companies waited it out and were driven to make adjustments through sheer economic necessity in a reactive way or face economic problems. Transforming business under pressure may not be the preferable way and certainly being forced to make business decisions is significantly different than being proactive in the face of the megatrends. For the regulated energy provider, this has often taken the form of being directed by the regulatory community to take action. In other cases, utilities find themselves pressured by exogenous events and have to scramble to meet their system needs, often at a higher expense in terms of dollars and reliability. It is clear, for example, that there will be some form of carbon tax and/or trading within the next 2-3 years—the only debate is the cost which ranges from $7 to $36 in current legislation. A reactive stance means waiting for the regulation/legislation to occur and provide surety, which raises the risk and cost of ameliorating the threat. Passivity in waiting for a megatrend such as the carbon/capacity challenge results in higher costs, less control, and greater uncertainty (Figure 2-2).

Other companies will make change through business strategies around the megatrends even though these investments in key areas such as core infrastructure, customer facing systems, and advanced technologies. Proactively transforming a business requires a

Reactive Business Transformation: *Economic Necessity*

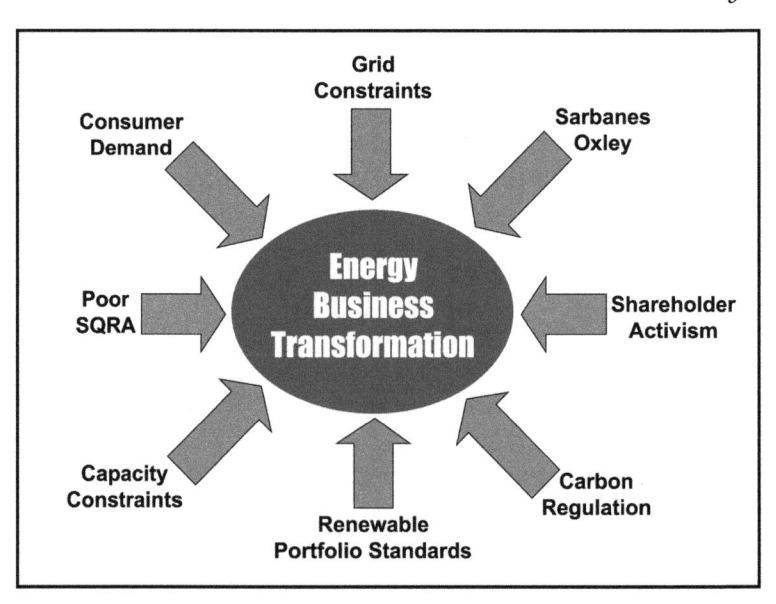

Arrows pointing to central "Energy Business Transformation":
- Grid Constraints
- Consumer Demand
- Sarbanes Oxley
- Poor SQRA
- Shareholder Activism
- Capacity Constraints
- Renewable Portfolio Standards
- Carbon Regulation

Figure 2-2. Achieving business transformation through a reactive process will be painful, expensive and arduous for companies and likely lead to failure.

different type of management style and vision as well as a higher risk tolerance. Understanding the megatrend of customer engagement has resulted in several large municipal utilities investing heavily in fiber rings and communications, often in advance of the market. Utilities such as KCP&L and Ameren made decisions nearly a decade ago to move into advanced metering, recognizing this would help in terms of intelligent infrastructure to relieve pressure on an aging workforce and to reduce costs. Proactive strategies require some visionary leadership as well as a leap of faith. However, it is easier to take a no regrets strategy when being proactive—and almost impossible to follow such a path in reactive mode. Just as building advanced infrastructure allowed telephony significant advances and the marketing of new services, so will similar investments in the utility sector (Figure 2-3).

In reality, it is likely that most companies will experience a

Proactive Business Transformation: *Megatrend Solutions*

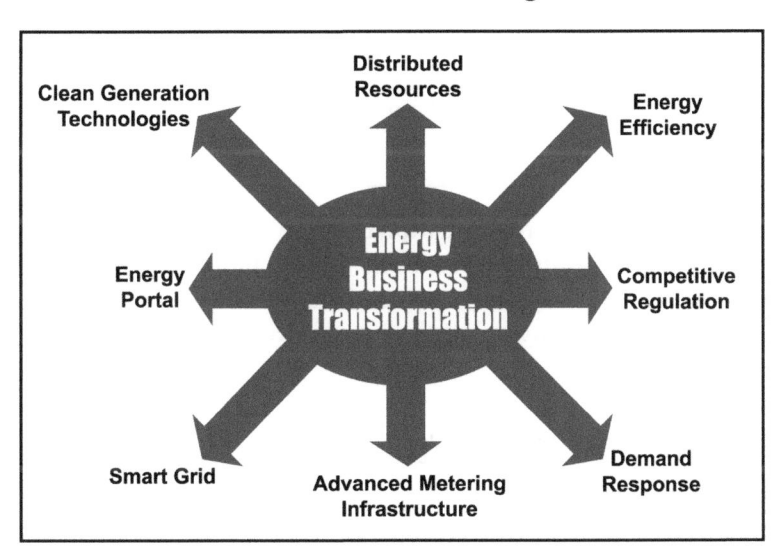

[Diagram: Energy Business Transformation at center, with arrows pointing outward to: Clean Generation Technologies, Distributed Resources, Energy Efficiency, Energy Portal, Competitive Regulation, Smart Grid, Advanced Metering Infrastructure, Demand Response]

Figure 2-3. Achieving business transformation through a proactive approach will be a challenging process of strategic planning around the megatrends, using a balanced corporate investment strategy yet will likely lead to success and business expansion.

combination of these two approaches, being reactive for those things that require huge risky investments where there is uncertainty of the outcome and the appearance of being a market trend and being proactive in other areas. (Figure 2-4). Understanding the megatrends and the distinction from market trends, will allow companies to find the right balance between these two choices.

Reactive and Proactive Business Transformation:
Economic Necessity and Megatrend Solutions

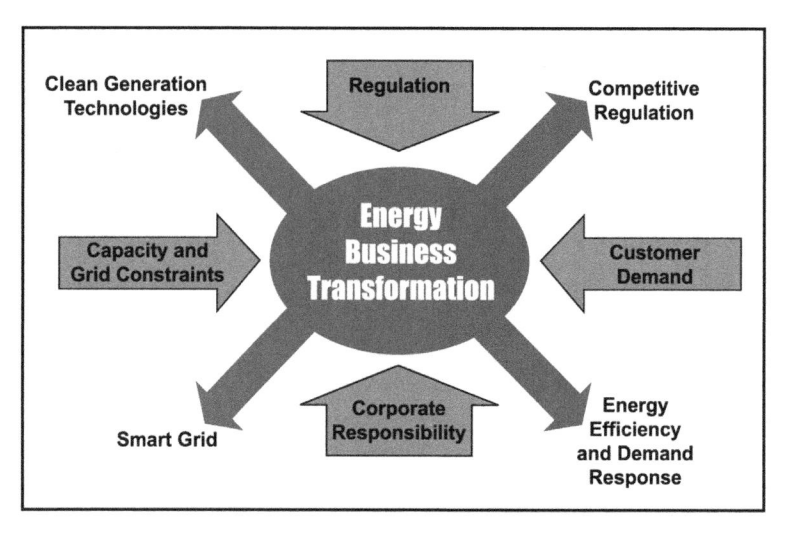

Figure 2-4. Business transformation through a combination of economic necessity and megatrend solutions is the most likely scenario for companies today. They feel most comfortable taking investment risk in those areas and waiting on other opportunities.

Competitive Regulation:
The Infrastructure Investment Model

As electricity companies return to a more traditional profile and attend to their core business, one might think that the competitive model is obsolete. But what the industry really needs is a new model for "competitive regulation" comprising features of both a

Incenting Innovation: A Competitive Regulatory Framework

regulated and competitive industry environment and balanced by a new corporate financial investment strategy. Significantly, such a framework will incent the innovation needed to build a sustainable electricity business for the future. As we've seen, no additional value-added services, successful or otherwise, can take the place of infrastructure investment and operations.

The Infrastructure Investment Model

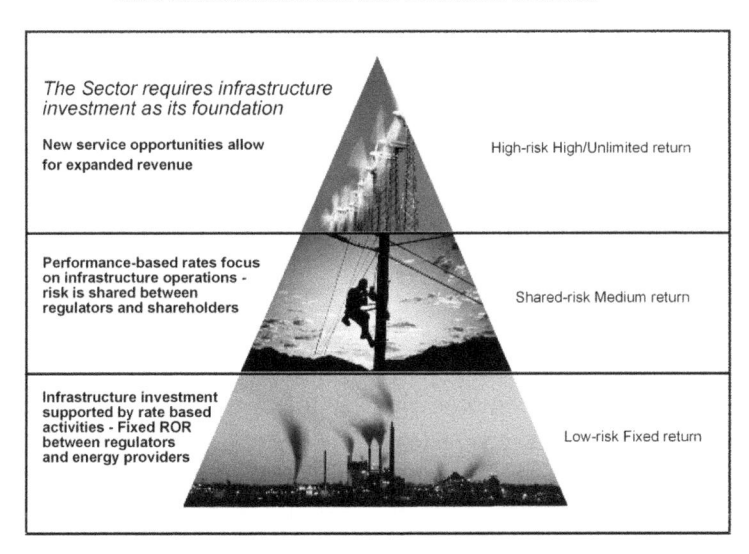

Figure 2-5. The industry needs a new model for "competitive regulation" which provides for a regulated component assuring fundamental electricity infrastructure, and a competitive component allowing for both performance-based and higher-risk venture investment.

The competitive regulation investment model, outlined in Figure 2-5, will allow for:

- A regulated rate of return on investment in new infrastructure supported by rate-based activities. This provides for solid investment in the necessary infrastructure needed to maintain a high standard of service and to launch new and valuable businesses for customers, individual companies, and the industry as a whole. Technologies that fit into this layer of the investment model are the Smart Grid, the consumer portal, and advanced, clean generation technologies.

- Performance-based rates for infrastructure maintenance and operations that provide a medium level of return and require risk-sharing between regulators and company shareholders. The more effectively and efficiently the company runs its business, the higher the profits and the happier the shareholders. Technologies that fit into this layer of the investment model are advanced metering and dispatch.

- And on the top layer, or the competitive side of the model, companies will still have the opportunity to invest in higher-risk ventures, such as new service opportunities allowing for expanded revenue and unlimited return.

This structure is used throughout this book to illustrate how technology solutions can be combined through different levels of regulation to help companies succeed with the various megatrends. The structure allows companies to open themselves up to a sustainable business cycle. First, the regulatory-consumer compact will provide a fixed return (through the rate base) on infrastructure innovations to ensure the health of the commodity business provision of basic services and customer care. Building on that foundation, companies can choose to invest in medium-risk, performance-based endeavors, as well as in higher-risk engagements to offer value-added services that enhance the basic service and provide for a higher possible return. This cycle is described in detail in the next chapter, The Destiny for Business Transformation.

References
Airline Deregulation, Alfred E. Kahn, *The Concise Encyclopedia of Economics*
White Paper: Is Innovation at an Electric Utility an Oxymoron? Jon T. Brock, COO UtiliPoint International.
Cost of Power Disturbances to Industrial and Digital Economy Companies, Primen Market Study completed for E2I/EPRI CEIDs Initiative, (Consortium for Electric Infrastructure to Support a Digital Society), June, 2001.

Chapter 3

The Destiny of Energy Business Evolution
A Megatrend of Reinvention

The scope and scale of projects, complexity of operating systems and the need for significant influxes of capital into the energy industry will continue to drive dramatic changes in the way business is done across the globe.

After 70 years of relative calm and stable models in energy, the changes over the last decade have set dramatic forces in motion that will require electric, gas and oil entities to continually reinvent themselves in order to stay competitive and in business. This holds true for public power as much as big oil, from building large-scale generating stations to interlinking distributed resources, and from managing bits and bytes across electrical systems to integrating hydrogen into the existing natural gas pipelines.

The staid model of vertically integrated utilities is gone forever as the investment community continues to push players toward economic efficiencies; the scale of projects increases dramatically in size and the financiers seek to continue deal streams. This is occurring down to the smallest player in the marketplaces as the availability of capital hinges on terms and conditions which require market model changes.

The delamination of both the electric and gas industries led to the creation of new entities which, in turn, developed new funding and financing models. Financial guarantees now require a level of complexity unheard of in the past, with the simple balance of debt-to-equity ratios overtaken by deals involving a whole host of new financial tools. The advent of new financial instruments and the creation of new companies to manage those instruments has changed the management of risk, stimulated change in models, as businesses are built around the mechanisms, not the hard assets as they once were.

Construction of billion-dollar power plants, the investment of tens of billions of dollars in advanced infrastructure, clean technology funds pushing the edge of generation resources, coal to liquids replacing traditional oil, all will require different business models. Other than the large investor-owned utilities (IOUs) with generation companies, few firms today can afford the construction of large power plants. The whole host of new players in independent power, independent transmission and retail marketing have changed the way in which the business is financed and managed. The concept of "unlocking value," driven by other businesses, has moved full force into the utility space.

As evidenced by the following examples, the trend toward business model evolution is clear and undeniable:

- In generation, the delamination meant the splitting off of power plant ownership from regulated business operations as utilities sought to get out from under regulation and regulators sought to "open up" the marketplace. In transmission, KKR's revolutionary move to segregate DTE's transmission system as a stand-alone entity resulted in a huge step up in value as an asset with a book value of $285 million sold for $600 million and was spun off in its own IPO at $1.3 billion—all without adding any fundamental value.

- Warren Buffet's acquisition of MidAmerican Energy followed by the takeover of PacifiCorp demonstrates the act of putting traditionally publicly owned companies into private hands and is another clear example of business model change. Buffet's strategy is "buy and hold" in this market space and he was able to placate state regulators through rate reductions and promises of financial stability over the long term.

- Kohlberg Kravis Roberts (KKR) attempted to purchase Tucson Electric Power and Portland General Electric from Enron. Both of these purchases were rebuffed by regulators (with threats of legislative action) who felt the KKR move (as the case with DTE's transmission system) was clearly a "Barbarian at the Grid" approach. The latest move by KKR, along with TPG, to take TXU private, in what will be the largest private buy-out in history (in excess of $45 billion), signals their continued willingness to pursue a business model unseen in prior years. The aggregation, disaggregation and reintegration signals a market in flux.

- Bill Gates, founder and chairman of Microsoft, already the largest shareholder in PNM Resources (9.6%), owner of Public Service New Mexico and Texas-New Mexico Power, has formed a multi-million joint venture with PNM known as Cascade Investments, looking to expand holdings in generation and other power sector assets.

- Purely financial buyers are looking at the industry—from co-ops to IOUs—as potential investments. Babcock and Brown Infrastructure, while rebuffed from buying Northwestern Energy, recently went to the market with an IPO aimed at raising capital for investment in the sector.

- Caprock Holdings, a formerly rural electric cooperative that was taken private by Lindsay Goldberg, an equity firm headquartered in New York City is buying Semco, a public company with holdings from Michigan to Alaska.

EPACT should result in an increased number of mergers by IOUs at a time when municipals and co-ops may be faced with the prospect of joining together to present a larger buying force to the market. Over time, it may be difficult to tell the traditional difference in utilities as partnerships are formed to own plants, IOUs and G&Ts work to support shared transmission infrastructure owned by third parties, and municipal utilities continue to offer power into the grid. Competition for services and staff means outsourcing, and the management structure to handle these changes will be different.

The megatrend of business model evolution means that companies will have to change and adapt in creative ways in order to survive and prosper. It will no longer simply be a case of adding an unregulated subsidiary, building out a product's program, or adding a fiber ring to a community.

Developing a Stable Business Model

Building a sustainable business model requires an understanding and planning for all megatrends as well as the management of existing and new market trends. It is important to note that market trends can change; megatrends will not.

In consumer electronics, companies (and individuals) got caught up in the market trend battle between Betamax and VHS. This is viewed as a battle that Sony (and a host of other electronics manufactures) "lost." There were certainly winners and losers in the short term market battle over the dominance of specification of the video recording medium. The reality is that the traditional players—Sony, Samsung, RCA etc. failed to see the entertainment megatrend of personal choice and spent time debating format while companies such as Apple with its iPod, TiVo with its digital video recorder and Motorola with its cellular phones that take pictures, play video and surf the web took advantage and secured positions in the marketplace. In the end the market trend of media supremacy led companies down the wrong path—the game changers took advantage of the megatrend.

The same has been the case in the electricity business as companies spent much of the 1990s and early 2000s preparing for the market trend of competition and deregulation. This chasing the tail of constantly moving re-regulation occurred across all spectrums of the business, from investor-owned utilities to municipal power systems. It is easy now, with some hindsight, to see that the market trend has been slowed or stopped and may never really take shape again in a tops down regulator/legislator model.

Those companies that have already anticipated the megatrend of business model evolution and the ensuing changes to the playing field are those who have already or will soon succeed. In the electricity business, it has meant delaminating vertical structures such as spinning off transmission from distribution, regulated from deregulated generation. Rather than trying to guess which way the regulatory rules would be made, successful companies revamped their business structures to be flexible enough to survive change—and to maximize benefits whether to shareowners, members or residents. Smart money has followed this megatrend, KKR betting on the value of DTE's transmission system, Direct TV aligning with rural electric cooperatives, municipalities breaking off from the Tennessee Valley Authority as costs continued to rise.

The financial community and Wall Street also play a significant, and often understated role, in working this megatrend. Stasis is not good for investment bankers as few fees are generated. Deals, refinancing, mergers and acquisitions all feed the megatrend and will continue to do so for decades to come. Sophisticated financial instruments have been brought over from other businesses and are now being deployed in the utility industry. The rules of the game are continuing to change for all players. Large municipals may find themselves challenged on bonding if they fail to be in compliance with requirements; co-ops similarly face revised risk profiles.

Energy Company Reinvention Strategies

The electricity industry has struggled with the transition to a competitive market. Viewing the examples of industries before them, companies felt empowered (or coerced) to take on strategies in advance of the impending deregulation to get a jump on the competition, protect their assets, impress their shareholders, and attract new investors. Many attempted high risk investments and reinventions that were not well suited for a commodity business such as electricity. Others were luckier, or more judicious, in that their investments eventually succeeded in their risk portfolios, broadening their base commodity and contributing to a reinvention of the company.

The freedom to experiment has been a stumbling block for many of the companies engaged in the provision of electricity. With regulatory boundaries fairly well defined, the current industry players moved around regulation rather than through it to a new end. Unlike telecommunications, electricity infrastructure has remained stagnant, preventing the more ambitious firms from attempting new business models based on technology and instead leaning on market maneuvers to create new business options. The challenge with relying on market alterations is that the change can be temporal; with technology it is changed forever—at least until built upon with the next technological advance. It is critical to recognize that electricity *is* rocket science built upon the laws of physics, not the law of the market.

Reviewing various trends in electricity company strategy over the last 10 years, it is clear that most companies transitioning out of the regulated environment are out of balance. There are five general strategies that companies typically adopted—from plain vanilla to risky business—to manage the challenges and opportunities offered by the then-advancing deregulation. Whether companies took a more traditional approach or tended towards a higher-risk strategy, very few escaped downturns in the market.

A more balanced business model—one based on a combination of the five strategies outlined—promises to protect companies from unknown market forces and provides for company reinvention. The robustness of this business model has been demonstrated in its successful application in several deregulated industries. In all cases, the business model has placed a high level of importance on technology innovation to its success. The future of the electricity business is also dependent upon technology investment and the argument for incentives is strong. We have laid out many opportunities for technology innovation in previous chapters, all aligned to one of four megatrends.

While a stable business model is the key to company reinvention, it will be imperative for the successful company to align with the destiny of energy business evolution and take on the opportunities available to them through investment. In order to take advantage of the megatrend and not be trapped in a market trend, companies need to position themselves in a new light. A review of some of the past attempts at following market trends provides some insight.

Plain Vanilla

Companies in the plain vanilla category focus clearly on a narrowly defined set of traditional activities, including commodity and service or pipes and wires (Figure 3-1). Seeking the safety of a known business, these companies fail to recognize that the known was changing underneath them. Trying to manage even a traditional business under a financially constrained model with no

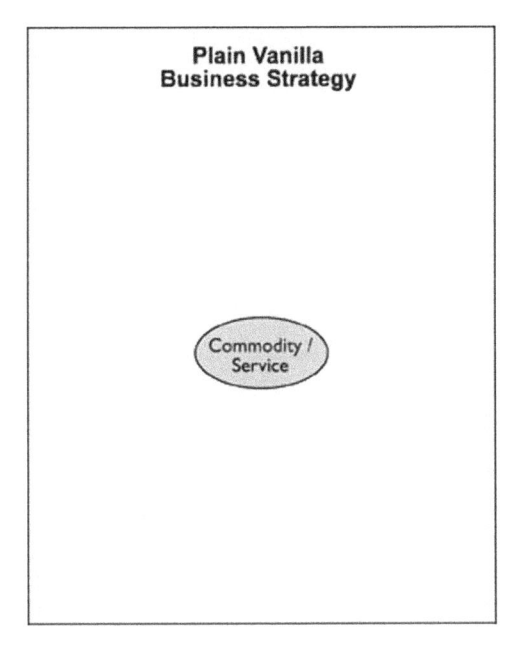

Figure 3-1. Companies with plain vanilla business models have maintained a safe and predictable, yet financially constrained strategy. While they have had some recent popularity due to stable stock prices, they will experience much greater challenges in the future due to a lack of investment in innovation.

rate relief (due to agreements with regulators) has proven as much of a challenge as trying to gamble with more speculative business models. Plain vanilla companies have not fared much better than companies that took a riskier profile. However, plain vanillas have at least maintained a repeatable and predictable level of business. Their stock, while not attracting investors, has been less impacted. Customer service has been negatively impacted as sacrifice to a rate-constrained world, although the quality of service has been maintained.

In current financial markets, plain vanilla companies are experiencing a renaissance of popularity—destined to be short-lived as markets move upward and system demands require significant investments that plain vanillas may be unwilling (or unable) to make. Though attractive in its simplicity, the plain vanilla approach will create much greater challenges in the future through its failure to integrate technology innovation.

Tech Savvy

Technology savvy companies—those that invest in technology for corporate advantage—are a rare breed. Some public power agencies and federal systems, having acquired the security of state-backed financing, have taken a leadership role in pursuing technology strategy. A number of IOUs, large municipals, and large co-ops have also moved in the technology direction.

Some tech savvy companies did not succeed because they viewed technology investment as a side business and did not merge it with their primary business (Figure 3-2). Several companies that pursued advanced metering stopped short of realizing the full promise of this technology because they attempted to justify technology risk only on the basis of a simple and rapid payback. With the right value proposition—one that made sense to regulators, investors, and customers—this technology risk could have been the foundation for reinventing the core business. Offering tremendous cost benefit and enabling customer involvement, it could have led the investing company to a new level of business.

Numerous companies actively decided to delegate—or

more appropriately, relegate—their technology investments to the venture side of their businesses under the guise of protecting any potential economic upside. These decisions created a divide between the part of the business that could appropriately apply the new technology and the business funding it. While these decisions were arguably a regulatory necessity, this strategy nonetheless left the new technology under-funded and out of the mainstream business operations, preventing its deployment and the possibility of a new business opportunity.

Few, if any, technologies thus managed have hit the big venture payoff or moved into the realm of expanding the core business. In fact, electric utilities of all flavors, including public power and municipal systems, have rarely capitalized on the potential benefits of their new technology investments. There is no one clear root

Tech Savvy Business Strategy

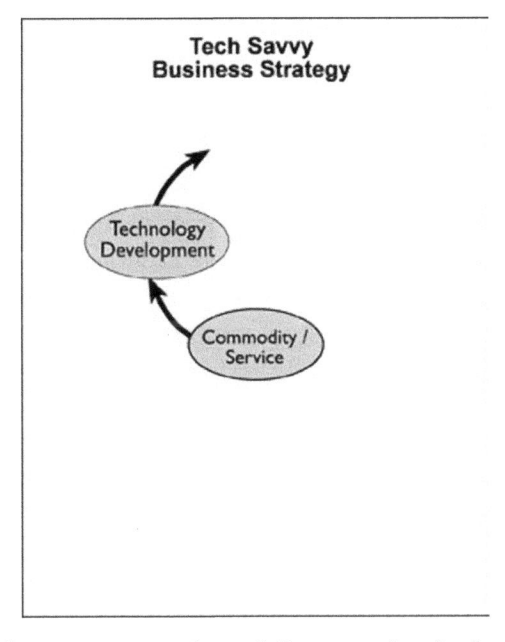

- Technology Development
- Commodity / Service

Figure 3-2. Tech savvy companies, while venturing in the right direction for achieving new business opportunity and company reinvention, have not succeeded in numbers because they did not fully commit long-term investment in technology ventures, minimizing business opportunity and longer-term benefits.

cause for the limited success of these strategies. Likely contributing factors could be the slow moving nature of technology adoption by the industry in general, a concern over taking too much risk, or simply the conservative wish to maintain the status quo in an engineering sense and prevent the adoption of technologies or procedures that could threaten the standard in place.

For example, tech savvy (and not so savvy) companies exhibited considerable foresight by investing in fiber optics to provide communications between their substations and other systems. However, they did not capitalize on the dark fiber they owned at a time when fiber was in short supply. Instead of providing high speed band-width to consumers—essentially developing and managing what could have been an end run around the baby Bells and the rest of the telephony world—the tech savvy engineering departments kept the fiber "dark" and "safe" at the expense of a new and exciting revenue stream.

Pac Man

Engagement in the merger and acquisition wars was a popular company strategy during the mid-nineties. The prevailing notion was that scale, scope, and size were critical for success. The mantra *"we must have more customers"* pushed companies into acquisitions that may or may have not been successful. In the end, this pac man strategy (Figure 3-3) has created combinations of companies that in many cases were not physically or philosophically aligned.

The pending repeal—expected by 2005—of the Public Utilities Holding Company Act (a gift from the 1935 congressional overreaction to our friend Sam Insull) means a host of new mergers will be eyed from those both inside and outside the industry. The question remains whether merging will create value for anyone other than investment banks or sufficiently reduce operating costs to justify the expense.

Consolidation in ownership of nuclear assets has clearly led to efficiencies and economies of scale in operations in the marketplace, and increased profitability on that side of the business. However, many Pac Man companies are facing performance challenges,

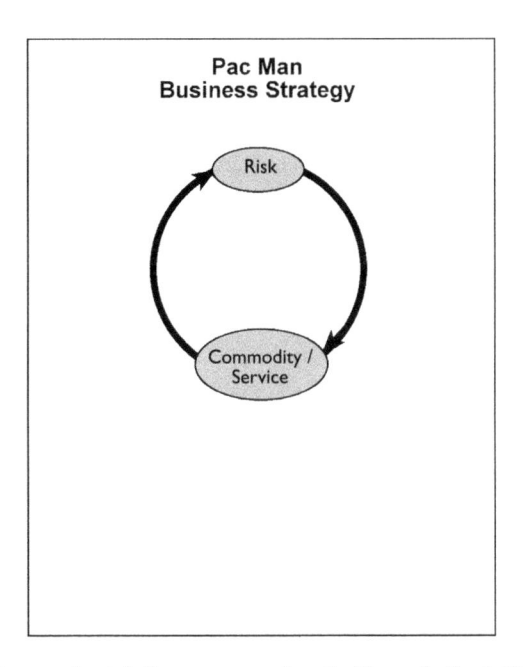

```
        Pac Man
    Business Strategy

         Risk
        ↗    ↘
        ↖    ↙
      Commodity /
       Service
```

Figure 3-3. Many electricity companies believed that the scope and size of a company were the critical elements for success, providing economies of scale and an abundance of customers. Unfortunately, large mergers created organizational challenges and distractions of their own, limiting business adjacency and business opportunity.

probably due in part to the distraction created by corporate mergers. Further, companies pursuing mergers and acquisitions as a growth strategy have been driven to work overseas with the idea of gaining experience in deregulated markets and of expanding their footprint in global markets. Thus far, there has been a significant lack of success—and several spectacular retrenchments.

Further, the lack of logic or business adjacency (competency, rather than physical, alignment) of many acquisitions prevented the realization of the much-anticipated boost in new business opportunity. Mergers are highly complex operations, and extensive planning is needed to ensure that they provide the expected benefits. In cases where synergy savings materialized, they often failed to expand profitability, reduce overall costs, or create the foundation for a new and sustainable business model.

Spinners

Another creative approach to deregulation was the practice of spinning off companies to achieve greater value. During the late 1990s and early 2000s, following the Wall Street mantra of "unlocking value," many firms divested various assets, hoping to match the double-digit return of an over-heated equity market. A standard procedure was to sell power plants while forming subsidiaries that acquired power plants elsewhere—avoiding regulation in a home market while "unlocking value" elsewhere (Figure 3-4). Companies also created new entities, such as independent power producers (IPPs) and energy service companies (ESCOs), to boost their market profile, raise capital, and sidestep regulatory burdens.

Lured by the irresistible vision of a balance sheet boost, spinners essentially hoped to have their cake and eat it too. In other words, by creating new entities in the marketplace, they aimed to remain in the base industry, increase attractiveness to investors and the value of company assets, and shake off the yoke of restrictive regulation.

Many spinner companies were spectacularly successful during late 1990s and early 2000s, primarily by placing generation assets into new operating entities, some of which were taken public on their own. For a while, high energy prices and high demand allowed these companies to sustain rates of return and maintain the illusion that even aging, sub-optimal plants could be profitable. Much of this exuberance was predicated on a cost of $2.50 per 1,000 Btu (MMBtu) for natural gas.

However, the high energy prices and demand that had forced plant prices higher and higher was temporary. The market forces had essentially masked true asset value and made even marginally attractive plants too expensive. As 2002 arrived, many companies became caught in the bust vortex of a boom bust cycle, facing lower demand, over supply, and the uncertainty of future contracts. Owners found themselves in negative financial positions as gas prices climbed and plant value declined. A number of companies went into bankruptcy, turned assets over to banks and limped through a power glut power with significantly overvalued plants.

The Destiny of Energy Business Evolution

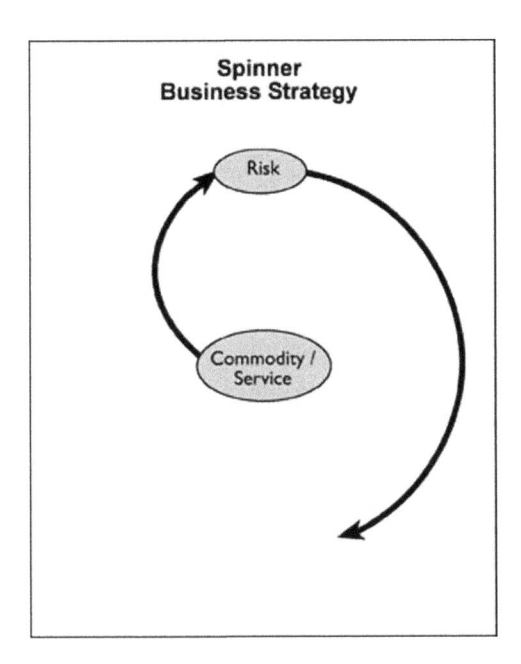

Spinner Business Strategy

Risk → Commodity / Service

Figure 3-4. The strategic practice of spinning off companies such as IPPs and ESCOs, was a good one as long as the market remained stable, unlocking value and attracting investors while avoiding regulation. Unfortunately, these spinner companies became caught in the 2002 bust market, and many failed financially.

In some parts of the country, the legacy of high priced plant purchases has led to the mothballing of facilities, impaired the credit worthiness of companies, and impacted the entire merchant plant sector.

In other arenas of the industry, the ESCO businesses also produced largely disappointing results. Independent ESCOs were purchased at inflated prices during the mid-1990s. Once utilities tired of these businesses, the ESCO spin-offs never lived up to their potential as demand for their products and services—driven in large part by utility rebate programs—evaporated. In the end, the ESCOs could not support themselves absent a large corporate parent with a demand for ESCO-like services.

The jury is still out on the independent transmission market as FERC rewrites rules for the industry. Some companies are now pursuing sale or placement of transmission assets into 3rd-party (spin-off) companies. The mid 2003, Kohlberg, Kravis, Roberts & Co of purchase Detroit Edison's transmission system as a private investment is an example of an aggressive new model that may unlock the potential value of transmission.

Risky Business

Risky business companies engaged in higher calculated-risk activities, such as trading, merchant power, telecom, and internet. These companies followed an asset-light or aggressive trading strategy in order to move beyond the traditional constraints and mentality of providing a public service (Figure 3-5). They have pushed the envelope to the edge of the financial cliff, where they could maximize securitization of assets, minimize actual financial investment, and plagiarize the script followed by Enron. The careful and calculated manipulation of accounting limits allowed for a high-risk, high-stakes, high-return game that required continued market expansion, continued price growth, and continued pursuit of deals.

Today, most risky business companies are challenged. Some are under investigation or facing legal action. Even companies that pursued a limited risk strategy are being unfairly tainted in the current view of their market position.

Companies shouldn't ignore the risk element of their business, as undertaking some risk is crucial to advancing to a new stage. Energy companies must be allowed to speculate on future costs if they are to be successful in anticipating market needs. In the past, power trading was not a means in itself, but permitted filling in for plant outages, weather, and backup supplies in times of demand. For example, like long-term contracts, weather derivatives are extremely valuable tools in balancing the cost of energy. The careful management of forward price curves, though risky by nature, provides opportunities for growth and profit and limits the downside exposure that electricity producers can afford to bear. If companies are not allowed to take risk, they will lose their ability to

**Risky Business
Business Strategy**

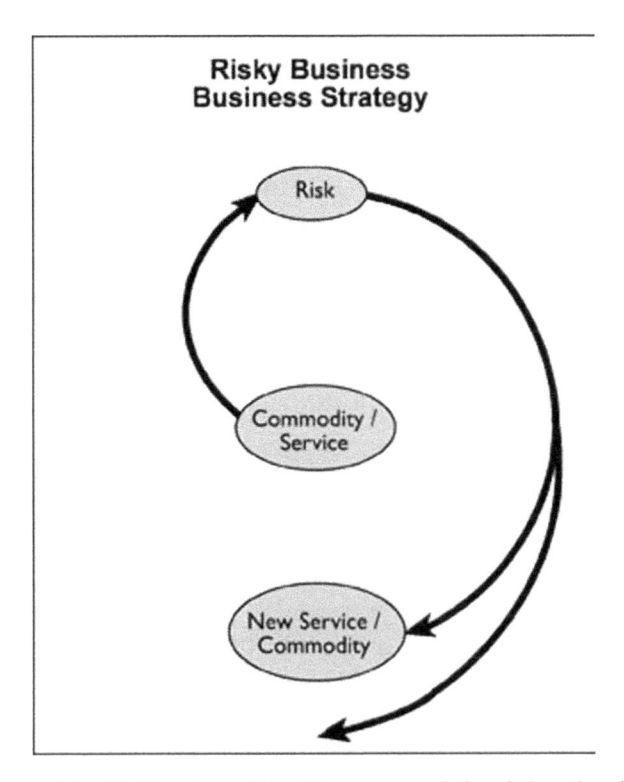

Risk → Commodity / Service → New Service / Commodity → Risk

Figure 3-5. Following the Enron model, risky business companies followed an asset-light, aggressive trading strategy entering into high-risk, high-stakes, and enjoying high returns. This strategy, while lucrative in the short term, requires continued market expansion, price growth, and pursuit of deals—an impractical long-term approach.

open markets and balance their business.

However, it is important to delineate and recognize levels of risk and the appropriateness of risk profile. The challenge today is to separate deals that are made for the sanctity of the system from those that are made specifically for profit. Regulators, shareholders and credit markets should have a right to be informed on a company's risk-trading profile and companies should be required to provide above-board information regarding their trading practices.

The Art of Business Reinvention: A Classic and Stable Business Model

A balanced strategy of the plain vanilla, tech savvy, and risky business company models, combined with acquisition and spin of activities that support the balance, is actually the recipe for a classic business model (Figure 3-6). In the right balance, these activities will lead to healthy and stable energy companies that will thrive in a competitive environment and stand the test of time. Healthy companies lead to a healthy industry, happy customers and an

Stable Business Model
Business Strategy

- Risk
- Technology Development
- Commodity / Service
- New Service / Commodity

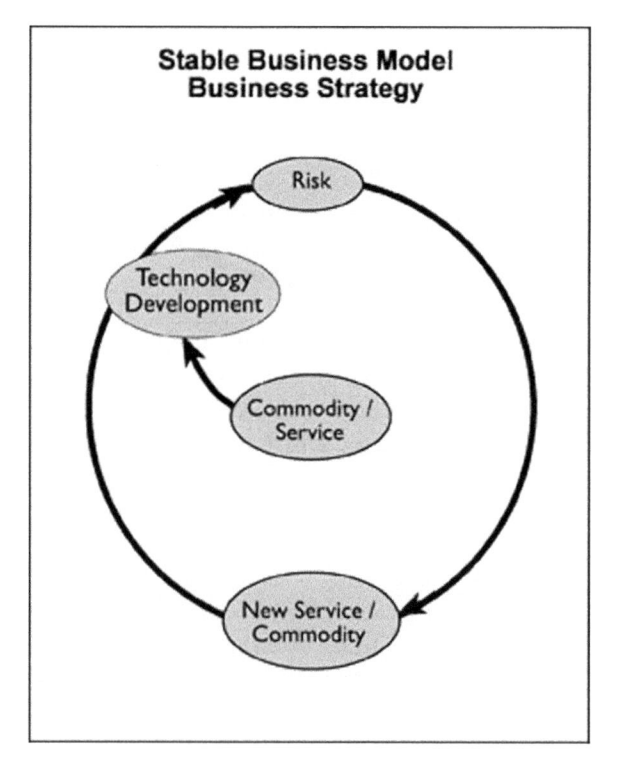

Figure 3-6. A balanced business strategy works in a cyclical process starting with investment in technology development related to the core business which provides for strategic and calculated market risk and ultimate reinvention of the company's core commodity/service business.

improved economic outlook.

The classic business model works in a cyclical process. All companies exist on their bread and butter business, their plain vanilla commodity or service that generates the bulk of their revenue. Profits from this stable revenue stream are used to feed new technology development within their core business. Without investment in R&D and new technology, the company will remain static. New technology naturally feeds the ability of companies to engage in strategic and calculated risk in the marketplace. This risk eventually creates new and expanded market share in existing and new markets. Through this expansion, the company has exercised its ability to reinvent itself with new commodity and service offerings. The additional profits from these new offerings fund further technology development and the reinvention cycle continues.

This balance in commodity, risk, and technology development is necessary to maintain a company's and an industry's ability to continually evolve and respond to market trends.

Examples of a Balanced Business Model

This classic business model can be seen in many prominent businesses today, as briefly described in Chapter 2. The following are more detailed examples of how companies blended technology investment and market risk to reinvent themselves, successfully building on their core business and maintaining market presence.

Airlines

Throughout the deregulation process, the airline industry invested heavily in enhancing operations, information systems, marketing, and pricing. These investments transformed the airline industry in ways that allowed it to deal with the free market system and streamline processes, and—for those that invested in and exploited these technologies—to remain profitable while maintaining customer service levels.

American Airlines had been flying people around the world in the same fashion for 30 years. They invested in new information technology, developing the Sabre Reservation System (Figure 3-7) at a cost of several hundred million dollars. Taking a market risk, they had their ticketing agents and travel agents use this system in booking flights, allowing American to streamline systems for load and seating management. The risk proved worthwhile. Sabre not only helped American build their commodity by selling more seats, but also became the industry standard as American sold the system to other airlines as basis for similar systems. The system has added to American's core business and is widely viewed as part of American Airlines' competitive advantage.

Continually among the most profitable airlines—and often the only profitable airline—Southwest Airlines has built its business on investing in technology to standardizing aircraft, maintenance, and procedures. To increase productivity and enhance operations, SWA

The American Airline Sabre System

Figure 3-7. When American Airlines invested in innovative IT systems to handle its reservation system, they were building the basis for much more—a successful system that streamlines processes and improves operational efficiencies. Sabre, now an independent company, has changed the face of the travel industry through the innovative investment and market risk taken by American Airlines. *Source: Sabre Systems, Inc.*

developed precision IT systems and software that helped them keep planes in the air, reinvent low-cost airline travel, and operate as best in class.

Telecommunications

Although there's been much negative discussion surrounding the effectiveness of telecommunications deregulation, there is no question that massive investment and successful implementation of new transformational technology has helped the communications industry increase revenues and markets by offering a dazzling variety of new products and services.

In fact, the deregulation of the telecommunications industry provided a catalyst for the commercialization of technology that already existed. In 1962, Bell Labs, then a part of AT&T, developed the first digitally multiplexed transmission of voice signals. The investment in this technology innovation and subsequent technology development resulted in a more economical, robust, and flexible network design for voice traffic. Taking a market risk after deregulation, phone companies used this technology to offer a host of new advanced network services such as 911 and 800 numbers, call waiting and caller ID. In addition, digital networking was the foundation for the convergence of computing and communications.

Nokia entered the telecommunication business as a cell phone manufacturer. Their investment in digital interface devices provided them with the technology to offer next-generation digital cell phones. They acted on the calculated risk that the market would move in the direction of internet-enabled cellular communications. It did, and Nokia profitted from this new commodity, reinventing themselves as a key player in wireless internet communications.

Auto Industry

In the auto industry, GM is a classic example of the business cycle. GM's traditional commodity business has consisted of cars and trucks, as well as auto maintenance service through their dealerships. They innovated by developing satellite communications networks, interlinking existing technology and infrastructure in IT and satellite telecommunications in new ways. Through this investment, they took a strategic market risk by developing and marketing the OnStar system (Figure 3-8), as well as innovative finance programs. In short, they developed businesses that not only connected them more closely with the customer, but took advantage of a known business model of steady revenue from subscription services. This has led to a reinvention of their business—a new commodity in cars with advanced communications systems and attractive financial packages.

The General Motors OnStar System

Figure 3-8. When General Motors committed to investment in satellite communications technology in order to expand their business, they built the driving force behind the successful OnStar system, now an industry standard in transportation communications and safety. *Source: General Motors, OnStar*

Retail

In evolving from a general store in Arkansas to a giant in the department store business, Wal-Mart has fundamentally redefined retail in the United States. That redefinition occurred not merely by opening more stores, but through investing in computer systems and advanced business processes. Wal-Mart took a huge market risk by implementing advanced IT systems for inventory and shipping operations and helped pioneer the application of advanced "smart" building technology, participating in demonstration projects that offered high energy efficiencies. And they were the first retailer that requested aggregated/disaggregated energy billing so they could manage energy use at a corporate level.

In addition to adopting creative technologies, Wal-Mart expanded its horizon by locating stores on the outskirts of smaller, growing communities and providing broader a offering: Wal-Mart is now the largest retailer of groceries in the country. Through this combination of strategies, Wal-Mart has successfully stretched its commodity business and become one of the leading retailers of low-cost items in the United States.

Entertainment

Apple Computer has successfully established and built a number of market trends in the face of the megatrend of personalized information and entertainment. From its onset as the revolutionary provider of simplified personal computers to the powerhouse it has become in entertainment provision, Apple has consistently reinvented its business and all of the businesses it touches. In large part, the company has understood the need for constant change in supporting an ever expanding personalized information and entertainment business. The company continues to build on its core products without fear of obsolescence.

While various manufacturers were competing over standards and the transition from tape to CD to digital formats, Apple was busily reinventing the business recognizing the megatrend was in personalized information. This meant not only hardware (the iPod) and software, but an entirely new business model in iTunes. The company's next foray, into telephony with the iPhone shows how it can take a basic platform and build an expansion business, always playing to the megatrend—and thereby setting the market trend.

The Unique Attributes of the Electricity Industry

Though examples from other industries are instructive, they cannot provide an exact template for electricity restructuring. In purpose, size, and complexity, electricity is an industry unlike any other. The continual supply of the right amount and quality of power to the right end user exactly when it is needed is absolutely essential to the nation's economic health and quality of life. When the electrons fail to flow, life as we know it shuts down. The crippling effects of power outages are a testament to electricity's essential contributions.

Further, as aptly described in EPRI's *Electricity Sector Framework for the Future*:

> "The U.S. power supply network is the largest, most complex machine ever created and engages the most complex enterprise. It involves some 5,000 corporate entities, 100 million customers, 4 distinct forms of ownership and multiple levels of regulatory oversight. This must function at all times with absolute balance in its supply and demand while trying to satisfy conflicting economic, social, political and environmental objectives."

Because of these unique attributes, some level of regulation will likely always be needed in the electricity business. One role of such regulation will be to uphold the time-honored concept of ensuring electricity access as a "public good." Since the framing of this concept early in the 20th century, electricity has become even more essential. There's every reason to expect that public good will remain very much at the heart of the regulatory-consumer compact well into the future.

In addition, regulation is needed to ensure the high level of cooperation and communication needed to keep the electrons flowing. Under deregulation, the industry's long tradition of ready communication was compromised, as companies turned to greater isolation as a means of creating competitive advantage. The problem has become severe, as reported in a DOE study of a series of major power outages in 1999. The study cited lack of communication between industry players as a major stumbling block preventing rapid resolution of the crises. The northeast power outage on August 14, 2003 (Figure 3-9), was further proof that communications and a united approach to the vitality and health of the U.S. electricity grid is essential to protecting businesses and consumers against further

AUGUST 14, 2003 NORTHEAST POWER OUTAGE: BEFORE AND AFTER

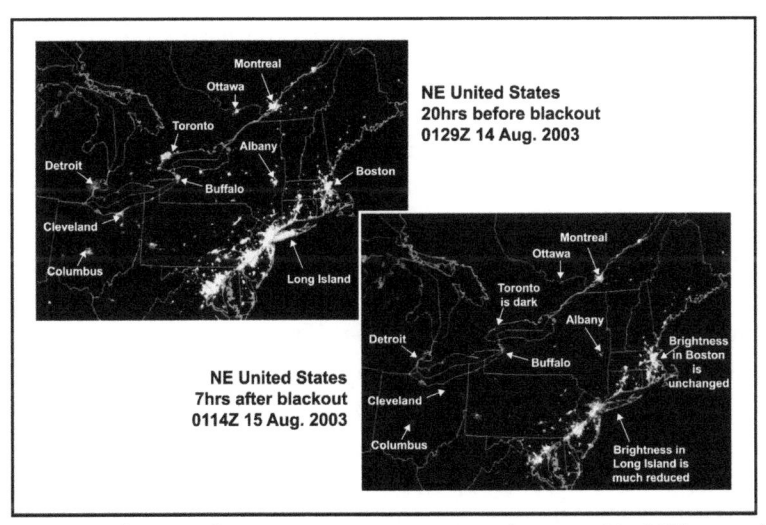

Figure 3-9. The Northeast power outage on August 14, 2003 was just one in a series of warnings that the electric power infrastructure was in dire need of upgrades. This outage, which cost businesses nearly $6 billion in the after hours of the day, also resulted from a lack of communications between companies. *Source: National Oceanic and Atmospheric Administration*

power failures.

Lastly, because of the grid's size and interconnectivity, the infrastructure enhancements needed to ensure adequate electricity service in the coming decades must take place on a macro scale. The efforts of any single company to improve a portion of the grid will of course be helpful—but only in the sense that "lighting a single candle" is helpful. Any meaningful attempt to improve the overall effectiveness and safety of the electricity infrastructure must be as vast as the grid itself. Simply put, developing the innovations needed to ensure the health of the electricity industry—and the economy that depends on the electricity infrastructure—is beyond the scope of any single company.

Calling for some level of regulation doesn't weaken the argument for competitive deregulation, a necessary and beneficial step. However, competitive deregulation will have to be structured to include provisions for public good access, inter-company communication, and collective investment on the scale needed to help the industry overcome the proven inadequacy of the current electricity infrastructure.

Incenting Technology Investment is the Challenge

The unfortunate thing for electricity companies today is that regulatory incentives to invest in their core technology infrastructure simply do not exist. Energy supply is a very complex enterprise, inevitably susceptible to becoming a political tool. In the old service model, it was always better to be 100 megawatts oversupplied rather than one kilowatt undersupplied. Today, the reverse is true, and the incentive for system improvement has been lost. Based on this pattern, the successful recovery of the electricity business depends on the incentives needed to maintain a robust, reliable, and efficient electricity infrastructure—one designed to keep pace with the needs of all consumers and the society they represent.

To fulfill the vision of a strong and sustainable energy sector, companies must maintain certainty and confidence through business models that reward leadership in innovation, promote a robust infrastructure, grow power supply and quality, promote realistic incentive mechanisms to resolve environmental conflicts, and strive to improve energy efficiency and electrification of the economy.

Perhaps the most visible impact of the regulatory quagmire has been the reaction of the financial markets. There is an inequitable higher cost of capital for electricity companies as the financial markets have punished the sector in general because of higher perceived risks. Investments in non-essential operations have come almost to a standstill. Due to the regulatory ambiguity surrounding transmission and distribution, most companies have restricted their investments until the market structure is understood more clearly. Investment in new power plants is almost completely limited to completing plants under construction or in paying penalties for canceled orders. Environmental investments in general are confined to only the absolute minimum that the current laws require.

Figure 3-10 shows the relationship between the IOU industry's annual construction expenditures and its depreciation charges. As shown, in recent years, depreciation expenses have exceeded construction, demonstrating that the industry is generally in a "harvest the assets" mode rather than an "invest in the future of the business" mode. Continuation of this trend has the potential to degrade the capacity and performance of the system. Due to an uncertain regulatory environment that disincents investment, and the pressures of cost containment over the last two decades, current infrastructure investment is now on the order of $20 billion per year, down from a peak of more than $35 billion per year in the mid 1980s.

This disincented investment scenario discourages energy companies from taking the calculated market risk that enables reinvention. It holds hostage a company's ability to adapt a classic, competitive business model and prevents the energy industry from moving forward technologically as well as financially. Electricity companies must address the causes behind investment disincentives to reverse the trend and develop a more classic, stable business and investment model that allows normal unrolling of the business cycle and continual reinvention.

IOU Construction Expenditures and Depreciation/Amortization Expense

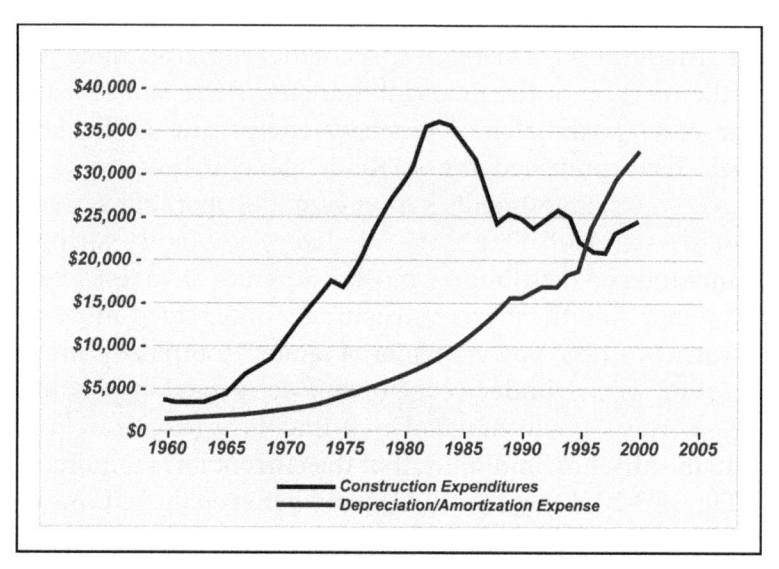

[Chart showing Construction Expenditures and Depreciation/Amortization Expense from 1960 to 2005, with values ranging from $0 to $40,000]

Figure 3-10. In recent years, depreciation expenses of Electric IOU's have exceeded construction costs putting the industry in a mode of "harvesting assets" rather than investing in the future. Continuation of this trend could degrade capacity, reliability and performance. *Source: EPRI Report Electricity Sector Framework for the Future, Volume I, August, 2003*

Succeeding with the Destiny of Business Evolution

It is easy in a business that has classically been regulated, either by state officials, municipally elected officials or its owners, to anchor around the status quo and point to the intransigence of those regulators as the rationale for avoiding or ignoring change. However, within the boundaries of this condition and those created by the physics of electricity, there are a host of opportunities for change—and a requirement for change in order to survive and prosper.

- Reinvention will mean changes in regulation and legislation—two areas in which utilities are classically very skilled. It will mean understanding and managing expectations of the publics being served in order to be successful. Regulation and legislation will change over time, and it will be critical to monitor and manage these changes.

- Ownership and governance will undoubtedly undergo significant alteration over time. Transmission systems will be increasingly owned by multiple stakeholders; massive generation projects will require governance models that have not heretofore existed, and distribution system management may even result in new models of operations.

- Customer choice aggregation and direct access will be driven not necessarily by regulation but by available technology. Technologies deployed for energy efficiency and energy management will result in new financial models and arrangements that give new power to the end-users. The aggregation of those customers may mean an end run around traditional regulation just as the internet has unified communities of users, cable has knocked the networks from their prime time perch and iPods continue to wipe out "album" purchases.

- Mergers and acquisitions will continue and may take different forms such as municipal assets being sold to IOUs, federal systems broken up and moved into the open market, co-ops joining together and independent renewable energy firms buy load management companies to provide a balance to their portfolios.

- Asset divestitures such as transmission spin-offs and utilities selling parts of their distribution systems will become more commonplace. Even for assets that are not sold in their entirety, a model of syndication will evolve around parts of systems.

Technology Opportunity Abounds

The good news is that a multitude of energy technology opportunities can bring companies into more strategic balance and pave the way for a robust power system of the future. Advanced technology can propel a company into future competitive success, although not always quick success. Advanced technical concepts for the energy industry, such as the hydrogen economy, the self-healing power grid, and advanced fuel cells, are among the visionary technologies needed to build a power system to support 21st century demands. Companies must consider both current and long-term issues—including the expanding demands of an increasingly digital society, an aging infrastructure, environmental challenges, and power quality requirements—as opportunities for technology development.

How can energy companies develop solid technology-based risk strategies that will help to reinvent the industry and prepare it for a sustainable energy future? What would the ideal energy company look like to succeed in today's business environment? Most energy companies provide a standard commodity of electric power and possibly a variety of value-added energy services. With strategic technology investment in the development of advanced concepts within the frame of competitive regulation, companies will be well prepared to take calculated risks in the market. The following chapters provide a host of possible technology investments that can fit into an overall strategy of the balanced business model.

The real challenge is not in finding opportunities to take advantage of the megatrends. The issue often lies in the willingness of management to take risks and stakeholder support of those risks. Rather than chasing the wisps of market trends, forward thinking companies place their bets on megatrends and adapt their businesses to forces.

Whether it is a municipal utility installing a fiber ring, a co-op pioneering Direct TV or an IOU taking a chance with new clean coal technology, the future is written _by_ those companies recognizing the megatrend and acting upon opportunity with calculated investment risk; history is written _about_ those companies who failed to see the future and take action in the business school case study of a failed investment strategy.

In order to succeed in this destiny, it is important to develop new products and services that revolve around the keys of SQRA—security, quality, reliability and availability recognizing this will require differentiated rates, levels of service and new financial models. A clear example is demand response as a true spinning which couples advanced technology with a new set of financial tools. This could very well be the Apple iTunes of the energy business which transfers control from traditional suppliers into the hands of consumers with a new set of middlemen who are only tangentially connected to the electricity business. This could rewrite the concept of rates and regulation, change the relationship with the market and alter a 100 year history of the provision of central station power. And, at the same time, this model change will open the doors to additional new participants.

This massive business model evolution could occur as traditional utilities invest in advanced metering infrastructure, capacity margins shrink, web tools continue their inexorable penetration in society and smart financial players invest in new opportunity.

As in any industry, the key to successfully navigating the destiny of business model evolution for the energy business is to try and write the new rules rather than follow rules written by others and find ways to deploy technology to advantage.

Chapter 4

The Destiny of Demographics
A Megatrend of Boomers

The destiny of demographics is a megatrend driven by the baby boom generation—the 78 million individuals born between 1946 and 1964 and a movement that has changed everything in its path since 1950. The first boomers turned 60 in 2006 and are beginning to retire in increasing numbers—and on their own individual terms.

This chapter breaks down the megatrend of demographics into three parts, yet the underlying cause is the same—an aging boomer generation. The issue involves the challenges felt throughout U.S. businesses in finding enough skilled workers to fill the ranks of the soon to be retiring. The megatrend also drives a migration of demand growth that will exacerbate the need for additional capacity in geographic regions already undergoing tremendous growth and the stretching of resources. And even as environmental consciousness becomes the norm, the boomer generation continues to drive consumption—consumption that has often been underestimated based on past forecasts.

The boomer bulge will have a substantial effect as it passes across American business, altering the patterns of the past in substantial and critical ways. While the issue of the aging workforce has been recognized and tracked for many years and is very much a reality now, the electric power, gas and oil industries are the most adversely affected by this issue. Large numbers of utility workers, engineers and managers are retiring in greater numbers than ever before while the need for skilled workers is at an all time high. Infrastructure development and environmental control will require an experienced and trained workforce over the next decade. Unfortunately, this reality is one that many power companies have done little about.

A number of business practices throughout the deregulation and restructuring era have exacerbated the effects of this trend. While dealing with an increase in raw numbers of retiring workers has clear solutions, over the last ten years, utilities have created a larger gap in skilled workers, particularly in technical and management positions through hiring freezes, layoffs and general downsizing of the workforce. Mergers and acquisitions have also taken their toll as synergistic savings were sought at the expense of succession planning. Utilities of all kinds are finding themselves with large gaps of institutional knowledge in some of the most critical areas of their business.

A significantly under-considered issue for many utility companies is the population shift that is just beginning to occur due to retiring boomers. 2006 heralded in an 18-year surge of retiring baby boomers, many of which will be moving to Sunbelt regions of the country where housing prices are more affordable and they can maintain their active, more affluent lifestyles. This movement will put further strain on already stretched electrical networks in these regions. Ironically, the very areas that are stretched today—Las Vegas, Phoenix, Albuquerque—are facing greater numbers of retirees to already overburdened systems.

Power providers must anticipate increased growth not only in sheer housing numbers but in increased demand due to larger homes, multiple homes and the increase of more and more sophisticated technology use as well. Baby boomers will not retire as their parents and grandparents did. Boomers are living larger, more active and longer lives.

Some numbers being projected show that residential energy use will go down over time; however, this may be underestimated. The energy use of electronic toys and equipment is on the rise, making up a larger percentage of the whole. While increased efficiencies in heating, cooling and water heating are expected to drive energy consumption down with an overall decrease of almost 10% from 2001-2030, the increase in electronic devices and the number of boomers owning second and even third homes in warmer climates may actually drive overall consumption upward.

Boomers are themselves largely responsible for the successful development and widespread use of technology in our society and can be expected to continue to be technology savvy, independent and active consumers as they grow older. The demand for high-quality power to build and maintain the electronically sophisticated lifestyles boomers are accustomed to will occur—power quality for technology such as advanced home entertainment systems, PCs, iPods, in-home offices, communications, and, as time marches on, advanced in-home healthcare and mobility technology. The surge in the demand for more, and higher quality energy is matched by the boomers' desire to have electricity provided in as clean a manner as possible. And there is an inherent contradiction in the boomer lifestyle of larger and sometimes multiple homes with more appliances while "demanding" green power, environmental convictions from industry and assuaging consciences by buying carbon credits.

Energy suppliers will have to learn how to manage the demands of aging customers. The aging boomer phenomenon, while anticipated to some degree in terms of market trend, has been widely misunderstood by power companies in terms of megatrend. The traditional practice of projecting regional populations with historical usage patterns will need to change. Types of energy use and time of use will change as boomers drive a new definition of retirement. Demands for more and higher quality power will continue.

As with all of the megatrends described in this book, it is important to recognize that not only will impacts be felt by power companies, but opportunities will be created. Planning models will have to be reexamined, companies will need to continue improving recruitment and training programs to fill the gap in service workers for the industry, technologies will be deployed such as fuel cells, not necessarily to power vehicles but to provide necessary power for the "toys" of life, and lifestyles will change to reflect power availability. Yet, of all the megatrends described in this book, the destiny of demographics will have the most profound effect on power companies and offer the most significant opportunities for success through investment in core businesses and expanded growth.

The Aging Workforce—A Gap of Knowledge

The aging workforce is not a new concept—it was predicted more than thirty years ago. The baby boomer population, those now aged 40 to 60, is made up of 78 million individuals representing 26 percent of the population and 44 percent of the total U.S. workforce. The U.S. population that is older than 55 will grow from 25.6 million in 1950 to 108 million by 2030. At least 30 percent of the existing U.S. workforce will be eligible for retirement in five years.

The next wave in the workforce—that of generation X and born between 1965 and 1979—is not large enough to replace the retiring baby-boom generation. This group only makes up 17 percent of the population and 34 percent of the workforce. Yet, these workers are in line to become the next group of leaders within U.S. industry. Attracting and retaining these young workers will be critical to utility success and will entail career-path movement that satisfies them.

There is no issue that will impact the electricity, oil and gas industry more over the next 20 years than the aging workforce and the ensuing shortage of trained management-level and technically skilled personnel. In fact, studies show that these industries will be the hardest hit by this megatrend. Historically, utility management staff has risen through the engineering and field ranks. Because of this, the boomer generation makes up the largest group of working-age individuals within utility management today.

At least 45 percent of the workforce in the electric and natural gas utility industry will reach retirement age within the next six to seven years and, by 2012, the utility sector will see a job shortage of 10,000 workers. In just five years, approximately 20 percent of these workers are expected to retire and while succession plans are in place for senior management, plans for technical and first-line managerial positions are limited.

This retirement boom will strain utility organizations throughout the industry. Solutions are compounded by current pressures to reduce staff at a time when an increased number of experienced workers are needed to meet the demands of infrastructure expansion, environmental controls and increasingly sophisticated power delivery system operations. Electric utilities will need to focus on increased efforts to attract new talent, train existing, more junior workers and retain their more experienced, skilled workers in creative ways.

The Booming Power Workforce Today

Mergers, cost reduction, rate freezes and a short-term focus on the bottom line has resulted in early and forced retirements, hiring curtailment as well as elimination of apprentice training programs.

Throughout the late 1990s and into the 2000s, utility business strategies were driven by cost reduction in preparation for deregulation. The most effective tactic to accomplish this came in the form of staff reduction. Investor owned utilities (IOUs), after making deals with utility commissions that froze rates for a set period of time in exchange for recovering stranded assets, were either legally or philosophically prevented from filing new rate cases. In an effort to both operate under a rate freeze and earn the allowed rate of return, many companies reduced staff, eliminated all costs where possible and reduced, if not gutted, new hiring. The short-term success of this strategy has now resulted in a lack of skilled workers available to fill the shoes of retiring boomers. This is coupled by the lack of interest of many young engineers to join a staid business with the perception of little head room. Add to this a lack of adequate training programs to mobilize sufficient staffing in the field and a megatrend of demographics emerges that will be a huge challenge for many utilities in coming years.

This megatrend will grow in severity over the next 10-15 years. As 2012 rolls around and utility workers start retiring in larger and larger numbers, the number of 35- to 44-year-olds—those normally expected to move into senior management ranks, will actually decline by 7 percent leaving a critical gap of corporate knowledge. In the same time frame, the group of workers ready to retire—those between 55 and 64 years old, is expected to experience the most growth—more than 50 percent. (Table 4-1).

Table 4-1. By 2012, ages in the labor force will change significantly as the group over 65, and set to retire, grow by 43% of the workforce and those entering the ranks of management, 35- to 44-year-olds, shrink by 7%. This will provide significant challenges for companies in efforts to protect their knowledge base and plan for the future.

Change in Labor by 2012		
Age	*Labor force change (thousands)*	*Percent Change*
16-24	20,110	9
25-34	32100	10
35-44	-24,930	-7
45-54	44,290	14
55-64	83,080	51
65 and older	19,410	43

Source: Occupational Outlook Quarterly

The labor shortage will affect many layers of the workforce, but will particularly impact engineering and field labor. Currently, the average age of utility workers is approaching 48 years while power delivery lineman average 52 years old and more than a quarter of all boilermakers are 50 years old or older. This represents the highest average age for any industry.

Virginia's coalfields have benefited in recent years from rebounding coal prices, and one problem currently facing the industry is an aging workforce, as the average age of a miner is now 50 years. Employment in coal mining has declined from 12,707 in 1989 to 5,288 in 2003.

Many power plants operating today have a very small and

aging cadre of workers with the inside knowledge of plant operations—knowledge that has yet to be captured and passed down to a new generation of employees. Even outsourced operating companies are struggling to build staff capable of managing complex systems. Additionally, half of nuclear-industry employees are more than 47 years old and as many as 23,000 retirements and other departures will likely occur during the next five years.

Attracting younger workers into an industry that is not known for its excitement, pay rates and chance for advancement is a huge challenge for power providers in general. Power plants, unlike other pieces of industrial equipment, have useful lives that continue for decades and entail highly complicated and critical operations. The location of some power plants has classically been problematic for attracting qualified workers as well—this issue will likely grow with the upcoming generation of workers that may be unwilling to move to remote locations for jobs perceived to offer very little future.

For public power utilities, the loss of critical knowledge and the inability to find replacements with utility-specific skills are the two biggest challenges as a result of the aging workforce. As is the case with other segments of the industry, a significant portion of public power employment base will be eligible to retire during the next five to seven years. The positions that will experience the most retirements may also be the toughest to replace and include first-line supervisors, senior managers and general managers.

Like their IOU counterparts, public power faces the challenge of attracting workers—yet, unlike their IOU counterparts, they cannot afford to pay the same rates or provide the same level of benefits due to the fact that municipal budgets, in particular, are being stretched on all fronts due to increasing labor costs and a smaller base of customers from which to obtain coverage. In some cases, linemen in public power companies can double their salaries by moving to a nearby investor owned utility or independent service organization. The security of a job with a muni or co-op may be no match for the lure of higher pay and profit sharing.

The solution to the aging workforce issue is far greater than simply finding workers to replace retiring boomers. Operational protocol and documentation have been put in place over the years in every level of the power industry. The old guard knows how the system runs, its history and the "tricks" necessary to keep it in order. As these skilled workers leave and with such a limited number of staff to pass this inside knowledge on to, the knowledge gap grows and operational efficiencies may suffer as a result. This is complicated by aging power plants and power infrastructure that need special attention and experienced know-how. Knowledge capture and process design are some of the tools being deployed to deal with the knowledge gap, but these are only part of the answer. Success lies at the heart of direct experience in how the equipment actually runs; the relationships developed over years of work experience and a broad understanding of the larger complex system.

Outsourcing—Solution or Cause?

Workforce levels in the electric industry have dropped by more than 23 percent since peaking in 1990 at 500,000 employees, even as power generation has increased by more than 30 percent. The movement to aggregate outside service providers in this arena began as utilities spun off their engineering groups and will continue this trend for the foreseeable future, further narrowing the field of candidates available to utilities. And, as system complexity grows, the demand for highly trained operators and installers similarly expands.

In a continuing trend starting in 1995, several large utilities, including TXU have outsourced significant portions of their operations. New Century Energies signed an exclusive outsourcing deal in 1995 with IBM Global Services worth approximately $500 million over the course of five years. NiSource recently signed a $1.6 billion, 10-year agreement with IBM and TXU outsourced its distribution workforce to InfrastuX.

The concept of outsourcing really took hold in 2001. Utilities, as they began to recover from the deregulation rollercoaster, began to reexamine their operations and cut excesses. Many had become financially stressed from poor investments in non-core businesses and energy trading operations. Outsourcing was seen as a way to improve their financial position without sacrificing operational efficiency. This led to the outsourcing of both technical and business operations. As companies focused on cost, and analyzed their strengths, any organizational process that could be handled by outside service providers better or more cheaply without sacrificing quality, were strong candidates for outsourcing. (Figure 4-1).

Utilities That Have Outsourced or Plan to Outsource in the Next Two years

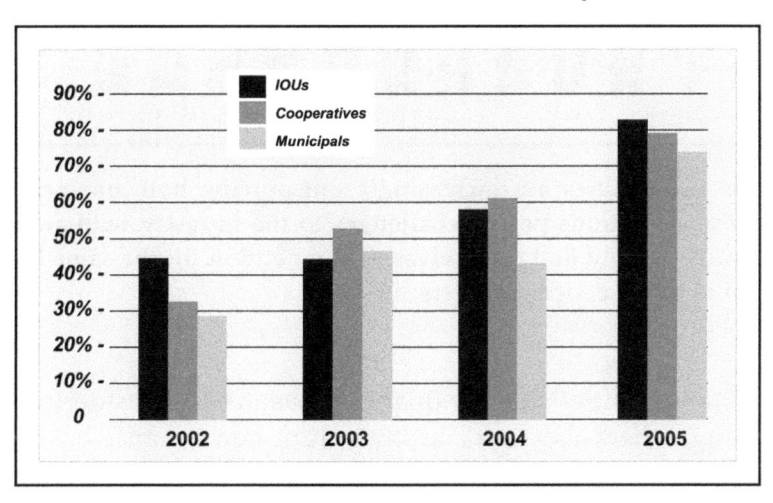

Figure 4-1. Companies are continuing the trend of outsourcing in both business and technical operations, helping them to improve financial positions, cut costs and maintain efficiencies. *Source: UtiliPoint International, Inc.*

Interest in outsourcing has continued to rise. At this time, a majority of utilities have moved at least a portion of their work to outside contractors in both technical and business capacities (Figure 4-2). These trends in outsourcing come at a time when companies are also increasing construction of new facilities and bolstering existing infrastructure with added intelligence—again creating the demand for highly skilled technical workers.

Utilities Considering Outsourcing

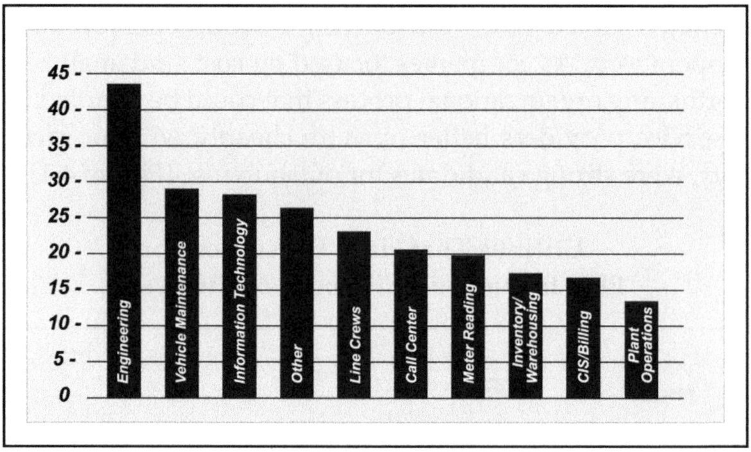

Figure 4-2. Utilities are increasingly outsourcing both business and technical operations posing challenges to the industry with an aging workforce—many find themselves in competition for the same human capital as their service providers. *Source: Black and Veatch*

Many utilities still cite financial reasons for considering outsourcing; others also now cite strategic or performance-based reasons. And this solution traditionally meets a utility's objectives. Yet, the practice of hiring contractors in an aging workforce represents a new challenge as independent service companies find themselves in competition for the same human capital as IOUs and public power.

Outsourcing is at once a potential solution and the cause of some of the hiring challenges. Utilities opting for outsourcing do so to reduce the costs and long-term expense of employees—yet as competition for workers increases, the prices they will have to pay also goes up. An irony of the outsourcing trend is that core utility competencies may be lost at an accelerated pace.

Though it will be difficult to outsource power generation in a traditional sense, off shoring of electrical equipment—and the skilled workers who build and design it—has already occurred such as in the transformer market. The industry stands at risk of its engineering and technical support functions being competed for across broad segments of related industries.

The Impending Knowledge Gap

In 2006, officers and managers of investor-owned and municipal utilities ranked reliability of their electric systems, electric infrastructure and aging employees as their three top concerns. As the number of experienced utility linemen decreases, the risk of outages increases. Linemen are the very ones responsible for maintaining the reliability of the grid. They are the workhorses who install and repair power lines, transformers and other gear and often must toil during storms, wildfires and other punishing conditions. A shortage of these workers could eventually limit utilities' ability to maintain or increase electricity supply, potentially affecting the economic and national security of the country.

The construction of new power generation and delivery infrastructure will require an assortment of engineering professionals—civil, electrical and mechanical. By 2010, the Bureau of Labor Statistics projects that the industry will need to expand its current engineer force by at least 15 percent. And the most extreme shortage will be in the category of utility field labor—electricians and boilermakers.

The power delivery infrastructure in the U.S. has suffered from a two-decade-long underinvestment while the demands being placed on it are increasing on a daily basis from the quantity of throughput demanded as well as security, quality, reliability and availability. Transmission investment, which has been lower over the past five years than in the Great Depression, is now projected to top $28 billion by the end of the decade—and is still estimated to be at 50% less than could be spent.

Encouraged by the 2005 Energy Policy Act (EPACT), utilities across North America and Europe are investing in technology to replace manually-read meters. U.S. utilities have installed more than 27 million remote-read meters out of a total market of 135 million and have announced plans to install another 30 million

Table 4-2. As the aging workforce impacts the number of available skilled engineers, lineman and other workers, the demand is growing. By 2010, the industry is expected to have a need for 15% additional engineers, a number that will be difficult to achieve within the current megatrend.

	2010 Labor Demand		
Skill Type	2003 Employment	2010 Employment	Percent Increase in Change
Civil Engineers	211,280	256,000	21
Electrical Engineers	149,540	175,000	17
Mechanical Engineers	214,070	251,000	17
Linemen	99,290	108,000	9
Electricians	575,980	819,000	42
Boilermakers	17,970	28,000	56
Construction Laborers	845,890	926,000	9

Source: Bureau of Labor Statistics

meters. Many other utilities are also considering their options as the technology improves and costs decline. While there will be less need for traditional meter readers, as this technology expands in use, utilities will require especially skilled field personnel to accomplish installation, maintenance and remote operations. In addition, there will be a huge service need based on technologies aimed at maintaining, improving and serving existing infrastructure such as system operations, performance optimization and asset management.

A major concern of the industry should be the impact that new construction and the aging workforce will have on the operations, maintenance and outage support for our nation's fleet of 103 operating nuclear units and 760,000 plus megawatts of fossil-fired generation.

Simple growth in this arena (2-3% per annum) calls for the construction of a new 300MW fossil power plant every three weeks and does not take into account planned retirements. Southern Company alone estimates it will need 5,000-6,000 workers to just complete its environmental control upgrades between now and 2012. Shaw Group's utility practice (formerly Duke Engineering Services) has grown from 150 to 1,000 in two years with the CEO offering that he would hire 200-250 more employees per year if he could find qualified workers—and Shaw is not the lead player in that market.

Plans around the globe to increase reliance on nuclear power face a potential stumbling block: a coming lack of know-how in running and regulating new plants. The reduction in the nuclear Navy and negative perception of nuclear power will leave that industry segment in particular straits just as it may find new plant orders coming after a 20-year hiatus.

The industry anticipates building 15 new reactors during the next decade. A tight labor market for nuclear skills likely won't stop a ramp-up in nuclear energy, however. Students and job applicants are beginning to respond to the potential for jobs and healthy paychecks in the nuclear arena, while companies like the nuclear business of General Electric Co. are increasing spending on training. In the U.S., nuclear-engineering undergraduates and post graduates have dramatically increased since 1999, according to a Department of Energy survey. Still, industry observers say a lack of skills and experience could add cost and complexity to nuclear ambitions.

The Shrinking Pipeline

Nearly 30% of the U.S. labor force with science and engineering degrees is older than 50. The number of bachelor's degrees in engineering and engineering technologies declined 2 percent between 1990 and 1996 and another 7 percent between 1996 and 2001. Other studies show a more dramatic decrease—colleges and universities reported a 50 percent drop in the numbers of graduating engineers in the last 15 years while degrees in computer and information sciences increased 74% from 1996 to 2001. Equally troubling is that university electrical engineering programs have shifted focus, eliminating or drastically reducing power engineering programs and putting the power education foundation at risk. Even federal government training through the Nuclear Navy is being curtailed just at the time when the industry needs it the most.

The ability of electric utilities to attract candidates is also a barrier to overcoming this megatrend. Over the last generation, the power industry has developed a less than glamorous image. Fewer and fewer young people are being attracted to the power industry, lured away to more exciting and lucrative high-tech industries. Many of the entry-level positions in the utility industry are not well suited for younger college graduates. While baby boomers were willing to sacrifice family for work, younger generations highly value and seek work-life balance. They are also much more attracted to organizations that are technologically savvy, team-oriented, and diverse. As many students graduating with EE degrees are choosing the more attractive high technology industries, increasingly, electric and gas utilities are competing with the oil and gas business for skilled workers.

While part of the solution will be to have more employees working later into their retirement years, this will not be possible in many physically demanding positions. Unlike some more traditional white collar jobs, there is an age factor after which some of the industry's most critical field workers will no longer be able to handle the physical requirements of these positions. In addition, attracting a new workforce has its challenges. A recent three-month advertising effort to attract linemen in the Florida market resulted in the hiring of a single lineman—and that at a 30% premium in pay.

About 57 percent of all utilities have a strategy in place for managing the empty pipeline and impeding shortage of qualified workers. However, that is considered insufficient considering that the positions that will experience the most retirements during the next five years are also those that will be the most difficult to replace—ranging from line workers to system operators to engineers.

Succeeding with the Aging Demographic

The loss of mature, skilled workers presents companies with a potential loss of institutional knowledge—a loss that can create organizational vulnerabilities and put a company at competitive risk. Yet, the significant demographic shift that is taking place also creates opportunities in the power market space. Entities serving the power needs of the nation and faced with this megatrend will be driven—whether proactively or in response—by the demographic momentum that is occurring and will have an ever-lasting role in reshaping the way power is used, generated and managed.

There are a number of strategies already being utilized by many utilities that will need to be expanded and adjusted to accommodate the coming worker gap. At the same time, there are opportunities for those companies willing to see forward, take risks and invest in a new future for their organization.

It will be imperative for power companies to recognize the aging workforce issue as in need of immediate and sufficient measures. Overall, companies rank the issue as relatively high in importance, yet not a crisis. Most prefer more traditional approaches such as hiring and training programs and don't consider tactics such as process improvements or knowledge management a likely strategy. Unfortunately, many of these companies will be ill-equipped to meet this crisis when it occurs in the next five to ten years. Approaches being used by some utilities now include:

- **Recruitment and Retainment Programs**
  Recruiting and retaining quality employees is an obvious starting point for companies putting together proactive strategies to deal with this megatrend. While many human resource departments have programs in place, more aggressive strategies will need to be taken to attract and maintain a quality workforce in coming years. Creative part- and flex-time options for workers will attract additional talent and encourage older workers to stay. The sooner the better; as time goes on, it will become more and more difficult to achieve.

Studies show that many of the younger workers coming into the workplace will be heterogeneous and female. These groups, in their 20s, need to be attracted into the industry with the right promise of culture and equal opportunity for advancement if companies want to retain them. For many utilities, this will require adjustments.

- **Employee Benefit Programs**
  Companies must also consider the value of maintaining workers, particularly older workers through creative retirement and benefit programs in order to reap the competitive benefits that their productivity, work ethic, skills, and knowledge can bring to their organizations. Benefits can include long-term care insurance, pre-retirement planning, health and wellness programs and prorated benefits for employees on flexible work schedules.

- **Training Programs**
  Most progress in addressing the aging issue in the workforce depends on effective training programs—for young worker career advancement and for technology adoptions. Organizations see a good return on investment when training is aligned with the company's strategic objectives. Many companies have training programs in place either within the organization or in partnership with others. In addition, some are also negotiating agreements with union locals to increase the percentage of apprenticeship program participants.

A number of utilities have partnered with universities and community colleges to help promote programs and internships for a variety of utility career paths. Colorado Springs Utility is teaming up with the University of Colorado to create power engineering classes as a part of the electrical engineering program. This not only positions the utility as a likely employer for many of the graduates, it trains students to fit the particular needs of their company.

- **Process Improvement**
  Refineries and power generators need to turn the aging workforce problem into opportunity by significantly changing the way business is conducted. Strategies continue to be developed by process companies in the wholesale energy industry that support process improvement and business innovation through the application of new technologies that require substantially fewer staff yet maintain the same energy production with greater levels of reliability and at the same or lower cost.

- **Knowledge Management**
  A largely under utilized tactic in the energy business, through knowledge management, companies can capture and document tacit knowledge from their workforce regarding critical processes. Knowledge management is increasingly being recognized by U.S. business executives as critical to their success, and for good reason. With expected retirements, critical intuitive or experiential knowledge must be captured and passed on to younger workers to maintain current efficiencies and avoid trial and error experiences.

- **Industry Technology**
  Younger workers may be attracted to utility jobs through technology. Many of the jobs in the field continue to lag behind other industries in technology adoption, yet digital applications in the power grid and in plant operations continue to develop and become increasingly sophisticated. Future high-tech jobs will be created with the advent of the Smart Grid, IGCC, advanced metering and other advanced power engineering applications. These can be exciting and challenging opportunities for up and coming engineers. As the industry evolves, these jobs may prove just as desirable and as the high-tech industry, providing an avenue into the utility universe and potentially alleviating some of the strain.

Yet, energy companies will need to implement outreach programs to overcome their current reputation as a low-tech and stogy working environment. The military ads of the 90s showing the travel, excitement, and education provided to enlisted young men and women come to mind. As discussed in other chapters of this book on infrastructure and power generation, advanced intelligent systems, clean coal power plants and renewable technologies may provide the impetus for increased interest in the business, as they will certainly provide increased financial opportunity.

- **Field Technology**
  Organizations also need to implement new field technology in order to fill the workforce and productivity gap. Advanced systems can dispatch workers to specific problem sites, system operations will allow more remote repairs and failure anticipation, substations with remote communications can direct power flow around problem situations, and fewer employees will be needed to monitor and manage power plants.

- **Outsourcing**
  Outsourcing is utilized by many utilities and will continue to be a significant strategy as they deal with a larger and larger knowledge gap. While not a long-term solution by any means, it may be an essential part of the workforce plan. As mentioned earlier in this chapter, outsourcing will present its own challenges as utilities and service providers compete for the same skilled workforce in some areas. Additionally, as outsourcing in the late 90s and early 2000s added to the current knowledge gap, it will only continue to perpetuate the problem as companies fail to build internal knowledge of their own. Consulting firms in the energy space find themselves equally challenge to staff up to supply the demand from their clients.

- **Succession Planning**
  Succession planning with an aging workforce will be key to or-

ganizational continuity and retaining critical knowledge. The individuals in the incoming generation of workers have very different educations, training and expectations than boomers. Organizations need to execute strategic planning and remain flexible in order to build a succession strategy that will support a company into the future. One of the major benefits of succession planning is that it forces an organization to formally assess its future needs. Issues such as identifying how jobs and the necessary skills to fill those jobs will change in the future and identifying the right candidates are critical to planning efforts. Preparing successors from within the company with adequate functional skills as well as the breadth of knowledge will also be key. Additionally, building adequate compensation and incentive plans to draw the right candidates and build integrated leadership will certainly impact the success of any plan.

These solutions are certainly helping to alleviate the growing knowledge gap in many utilities. Results vary based on a utility's needs and strategy and success will depend on how seriously utilities are taking this issue and how soon they begin implementing adequate measures.

As we have defined in Chapters 2 and 3, utilities must leverage their core business with calculated business risk to make technology related investments and spur company reinvention. In this way, megatrends, such as the destiny of demographics, can be used to their advantage.

The provision of engineering and construction services is an area where incumbent utilities can generate new streams of revenue as well as business opportunities. While there will be significant efforts in outsourcing services, those utilities that maintain and build in-house capabilities will be able to use those strengths in a tight labor market. Other utilities with in sufficient human resources will be forced to pay higher rates for manpower to those firms that expand and market their staffs. The strategic benefit may be to allow an expansion into new markets and can play a role in merger and acquisition space as well.

The development of process improvements through technology is another area where utilities can find significant operational benefits in the face of a declining workforce. Where today it takes staff in trucks and on foot to perform read-in/read-outs of meters, tomorrow's systems, coupled with process changes, will result in few people handling the workload. One city in northern Florida reduced its truck rolls for such services by 40,000 every year, saving human resources, environmental impacts, and more than a million dollars in expense—not to mention wear and tear on vehicles and roads.

A major challenge for utilities will be environmental control installations, as the workforce available to manage and perform such projects is limited and in high demand. While utilities in the past had their own construction crews to handle such work, over the years those pieces of many organizations have been dismantled. For utilities who still maintain such capacity, there is an opportunity to provide services.

The monitoring and capture of imbedded knowledge and dissemination to the remaining workforce is a challenge and opportunity for utilities. As the brain drain continues over the course of the next 15 years, those companies which use tools and techniques to grab and secure the information will succeed nicely and may be able to apply it to other firms.

All of this requires significant levels of cultural change as well as investment in new businesses, process improvement, technology and training, if this industry is expected to prosper. The power of the industry to take charge of its own path is strong, providing the willingness to make the investments in time and technology. Ironically, many of the new opportunities in the future power industry will not be addressed by power providers at all, but by entrepreneurs capitalizing upon an industry ripe with market opportunities.

Other industries have seen similar phenomenon by following market trends but failing to see the bigger megatrend at work. AT&T made a conscious decision in the 1980s to exit the cellular telephony space, only to buy it back less than five years later at a cost of $3 billion; IBM ceded the PC space twice in the past 20 years and is now struggling to understand why its market penetration is slipping; U.S. auto manufacturers have consistently misread the marketplace, leaving profitable businesses to newcomers and competitors who then turn and encroach further on the remaining business market.

While utilities are busily getting their financial houses in order, they cannot lose sight of the elephant in the room—the destiny of demographics. The industry must be proactive. Investments in current and future employees pay off by increasing loyalty and providing the talent to organizations to meet their ever-changing needs. Investments in businesses that support the core function of the company and attract qualified employees build the corporate portfolio and improve revenue. The marketplace must eventually respond to impending labor shortages to ensure the transfer of knowledge to the next generation of workers. And although a seamless transition clearly is not guaranteed, those companies that work with the destiny of demographics have a greater chance of succeeding with the megatrend; those that follow market trends or fight the megatrend will likely fail.

Growing Population Shifts— A Booming Retirement Issue

Baby boomers have been trendsetters throughout their lives and, as they retire, this isn't expected to change. 2006 was the beginning of an 18-year demographic wave in which approximately 4 million people—20 percent more than in previous years—will have the opportunity to leave their full-time jobs and either stay put or purchase a retirement home somewhere else. Economists predict that at least 400,000 boomers a year will choose greener pastures beyond their state borders.

Aging boomers already are opting for unconventional, more rural retirement locations, primarily in the South and West. A survey of boomers indicates that 42 percent of boomers would like to retire in the South, 32% in the West, 15 percent in the Midwest and 12% in the Northeast. This tells us that the Sunbelt will remain a traditional draw for retirees. In the West, scenic towns are increasingly drawing active boomers, many of whom are retiring early and want to work part-time.

A significant portion of baby boomers married later in life and had children at a later age, which means many will continue to work beyond the traditional retirement age. Most boomers expect to work well into their 60s, and 27% say they never intend to stop working. Older boomers are thinking about retirement, but one-third expect to go back and forth between periods of work and periods of leisure, and another 35% want to work at least part-time or start a business. All of this will have an impact on the kinds of homes they buy as well as where they buy them.

As boomers are working longer, this has resulted in a significant migration of older urban professionals to nearby small towns, allowing them to stay close to business contacts and adult children while establishing a new life in quieter surroundings. This explains how Boston area residents have moved in large numbers to small towns in New Hampshire and an increasing flow of boomers to small towns that have cropped up around Atlanta. Another migration of city-weary and equity-rich boomers is occurring between high-priced California cities and communities surrounding Las Vegas and Reno, Nevada.

Recent population growth has been greatest in the suburbs. The urban cities in metropolitan areas averaged growth of 3.5 percent in the 1990s, while suburban areas outside the central cities climbed 12.5 percent. Overall, Las Vegas is the fastest growing area in the nation in terms of growth rate; however, the Phoenix area has experienced the largest influx of population in the country according to the 2000 Census 2000. (Table 4-3)

The migration to warmer climates continues to put pressure on existing power systems already straining under rapid growth due to the desirability of climate, business growth and the concomitant influx of immigrants to service population needs. As greater numbers of baby boomer Americans migrate to the South and Southwest, the

Table 4-3. A growing number of baby boomers are retiring primarily to suburbs in warm climates, putting increased pressure on electricity systems in those hardest hit areas. While many will continue to work part time, they want to be near urban centers where professional contacts remain, but enjoy the benefits of larger more suburban or rural homes nearby. Hardest hit is the Las Vegas area with the fastest percentage growth and the Phoenix area with the largest influx of population in the country.

Growth Rate of U.S. Metropolitan Areas from 1990 to 2000

Metro Area	Growth	Population
1. Las Vegas	83.3%	1,563,282
2. Naples, Fla.	65.3%	251,377
3. Yuma, Ariz.	49.7%	160,026
4. McAllen-Edinburg-Mission, Texas	48.5%	569,463
5. Austin-San Marcos, Texas	47.7%	1,249,763
6. Fayetteville-Springdale-Rogers, Ark.	47.5%	311,121
7. Boise City, Idaho	46.1%	432,345
8. Phoenix-Mesa, Ariz.	45.3%	3,251,876
9. Laredo, Texas	44.9%	193,117
10. Provo-Orem, Utah	39.8%	368,536

Source: U.S. Census Bureau

demands on the network for supply of energy will continue to increase beyond the "normal" projected growth patterns.

Evolving Infrastructure Planning

The shifting of populations has classically caused dynamic changes in infrastructure, whether it is the need for more roadways, more schools or community services. The shifting of population has already caused and will continue to create challenges for U.S. power suppliers.

The underpinning of electric and gas infrastructure that utilities are struggling to keep up in expanding parts of the nation will be stretched even further as the boomer generation flexes its muscle. Development is occurring in places where the architects of the local power systems never imagined, with loads never envisioned. Boomer populations are demanding greener power without the desire to see additional transmission or generation constructed. This perspective, while derided by some in the power business, requires utilities to think, invest and act differently. The fallback position of regulatory protection may not be enough to counter the pressure of boomer demands on issues ranging from undergrounding and environmentally friendly power generation to power quality and high reliability.

Resource planning has already undergone significant change and will continue to do so as the traditional business models shift along with populations. The paradigm that has been in place for more than 50 years is changing as evidenced by regional views of supply and demand, the need for better real time load forecasting and planning and the drivers of populations who will demand better prices for load control flexibility.

Technology advances will certainly play a role in securing grid operations as well as allowing flexibility and choice for customers with an increasing appetite for high quality power as well as the desire to manage and control their environments. The intelligence necessary to allow for such customer interaction means opportunities as well as challenges to the incumbents supplying power. Those areas without open retail access face a more demanding public used to getting their products and services in a customized fashion. While this may be viewed as a heresy by some, nevertheless, consumers will want to "have it their way."

The challenge will be to meet the new customer demands while squeezing more capacity from the existing system. Rising expectations will put pressure on incumbent, traditional suppliers; rising consciousness will mean companies will have to be responsive with cleaner energy—and more of it. As with all megatrends and the market trends that support them, this will be a boon to some, a boondoggle to others, and an anathema to traditionalists.

Increased Consumer Demand— A Boomer Phenomenon

The baby boom bulge not only creates a shortage of workers and population shifts to warmer climates, but an increase in overall demand as well. Unlike past generations when retirees classically downsized and simplified, the tendency of the baby boom generation is doing more with more. From larger living spaces and multiple homes to an influx of electronic toys and devices, larger-than-ever entertainment systems, home offices and in-home healthcare, the 78 million potentially retired, or semi-retired, boomers in 2030 will be living longer, living better and requiring additional power to support their needs.

The senior housing market is transforming the way builders design and develop active-adult homes including everything that goes into equipping the home. Boomers love to socialize—they want that large gourmet kitchen and plenty of room for entertaining. They want the larger master suite and the Jacuzzi tub—all requiring extra power.

In the past, technologies associated with aging were generally applied to disease or disability. Today, however, aging demographics and baby boomer lifestyles seem to be calling for a different kind of technology. The approaching wave of wealthier, healthier and more independent boomers is reframing the discussion of technology needs—and this is not just about charging additional golf carts. In response, technology and public policy will reflect greater emphasis on extending independence, productivity, and quality of life for the nation's senior citizens.

More boomers will work later in life with a significant percentage believing they will never fully retire—out of desire as well as necessity. The shift, however, will evolve current trends to the home office/small office model that has already changed the dynamic of the way in which we work. The demands for power reliability and power quality in the home will continue to rise.

The House Electric: Boomers and their Toys

Energy consumption in the average American home rose nine percent from 1987 to 2001 after accounting for energy efficiency improvements. Almost all of the growth was due to lighting, appliances and other electronic items which together accounted for about 30 percent of home energy use—things such as home entertainment, computers, and electric toothbrushes—basically everything other than space and water heating, cooling, and refrigeration.

Boomers, who are better off financially, far more active and have greater quality of life demands, will continue to drive technology use and energy consumption. The Energy Information Administration (EIA) projects an 8% increase in energy demand from these same loads by 2030 (Figure 4-3). Other studies indicate a 100% increase in electricity demand from ICT devices, or information and communications technologies, particularly home entertainment and large screen TVs, traditional boomer toys.

Even with appliances becoming more energy efficient, housing more insulated, and the population having a greater consciousness of environmental impacts, the number of boomers using more technology in greater numbers means additional electrical load and increased stress on the system (Figure 4-4). Retired boomers' homes do and will have increasingly larger home entertainment systems. A TV in 1984 averaged 24 inches and 42 watts of power, today it averages 42 inches and 240 watts plus 116 watts in standby power, and is accompanied by many other pieces of equipment such as DVD players, cable boxes, subwoofers, etc. In addition to this, many boomers have more elaborate computer systems and multiple homes as well, increasing demand even further.

More boomers will work through retirement, at least part time, so usage for in-home offices will increase. Their home offices will have all the latest computers and related gear requiring higher power quality and higher power reliability. Additionally, boom-

Delivered Residential Energy Consumption by End Use, 2001, 2004, 2015 and 2030
(Million Btu per household)

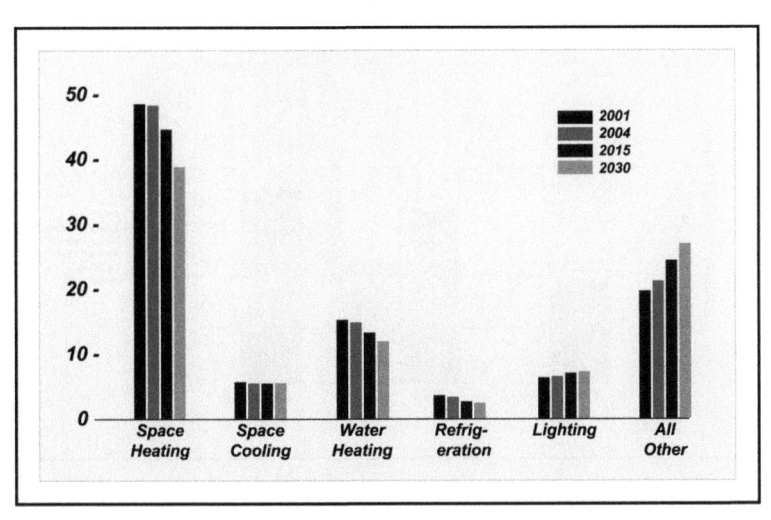

Figure 4-3. The fastest growing electric energy use in the home is that of "other" which is growing primarily by information and communications technologies (ICT), and made up mostly of more and larger entertainment and computer systems, traditional boomer toys. *Source: 2006 Energy Outlook, U.S. Energy Information Administration*

ers are more likely to adapt to new technologies such as iPods and wireless devices further increasing loads. And as plug-in hybrid vehicles hit the market within the next decade, boomers, as well as younger generations, will adopt this electric load as well.

In certain growing boomer retirement regions, new and higher consumptive technologies such as space cooling systems will be phased in. In New Mexico, an unexpected shift to air conditioners from swamp coolers has challenged energy supply. Ironically, the shift was stimulated in part by the City of Albuquerque incenting residents to switch to save water!

Boomers will demand continued independence and mobility into their later years. The significant increase in healthcare equip-

Total Residential Electricity Consumption for 1980, 2005 and 2015 (Projected)

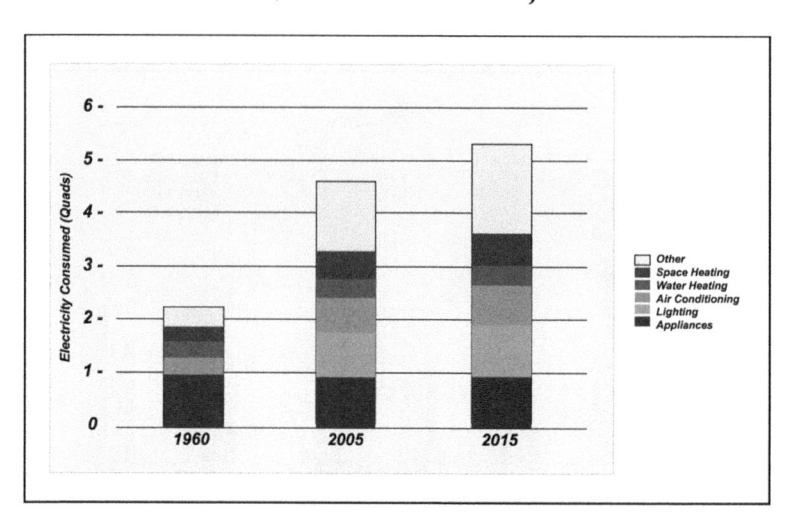

Figure 4-4. The DOE's Energy Information Administration (EIA) provides further evidence of the growing and projected increase in home energy consumption from "other" uses, primarily home electronics. By 2015, at least 30% of home energy use will come from this category. Sources: 1980 data—EIA's Building and Energy in the 1980s, June 1995, Sanchez et al., 1998. Note: In 1980, EIA grouped lighting into the appliances category. Lighting has since been broken out separately as its own category, as reflected in the data for 2005 and 2015. 2005 and 2015 data—EIA Annual Energy Outlook 2005.

ment that will migrate from hospitals and care facilities into the home is an additional part of the equation. This equipment will need special circuitry and a certain requirement for a higher level of power quality. This does not just substitute hospital equipment for home, it will represent a significant increase in the quantity of heath care equipment functioning in the home and servicing a single person which will require more energy than hospital units servicing multiple patients. More mobility devices will also mean higher power consumption.

History has proven this generation will change the dynamic as it has over and over again. In the 1950s the thought of multiple televisions would have been considered strange; in the 1970s the idea that there would be a computer in the home (much less multiple systems) would never have even been considered; in the 1990s the idea that family members would own multiple cell phones or pocket-sized devices to play music, videos, and books on tape was unthinkable. As time marches on, the types of devices and the sheer number in the home continues to increase, particularly electronics and media, along with the power necessary to drive them (Figure 4-5 and Table 4-4).

While individual appliances such as refrigerators are growing more efficient, the raw numbers are also growing as is the size of the units. Ceiling fans sprout in every room, even outside on porches; higher-end homes feature refrigerators and cook tops in

Penetration of Media Devices in U.S. Homes

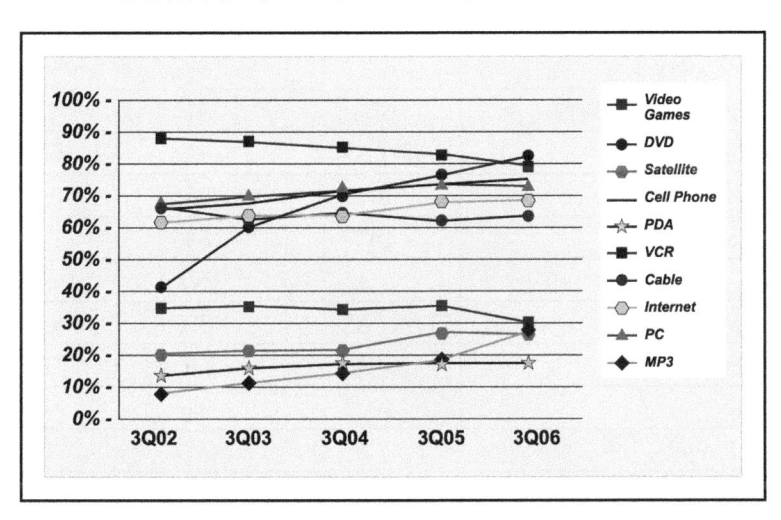

Figure 4-5. The growth of electronic media devices in the U.S. home is increasing at alarming rates. While devices such as VCRs are on the decline, newer technology such as DVD players and MP3 players are growing in popularity and use, increasing energy use in most U.S. homes. *Source: Nielsen Media Research's Home Technology Report*

Table 4-4. The number of different appliances and gadgets in the home are on the rise. While some of these items are small, their sheer number can increase energy use substantially over growing populations. The Rocky Mountain Institute found that in an average American home, energy use rose 9% from 1987 to 2001, almost all as a result of lighting, appliances and other electronic items equaling 30% of home energy use. § Numbers have not changed appreciably or not available.

Gadgets for Every Room

Electric Device or Appliance	Percentage of Homes with each device		
	1987	2001	2006
Microwave Ovens	66	96	§
Clothes washers	70	94	§
Clothes dryers	60	78	§
Freezers (standalone)	41	47	§
Blenders	64	82	§
Dishwashers	48	59	§
Electric coffee makers	67	81	§
Cordless phones	N/A	81	83
Answering machines	10	78	§
Computers	15	63	73.4
PDAs	N/A	N/A	16.4
Printers	N/A	57	§
Modems	N/A	56	73
VCRs	52	94	79
DVD Players	N/A	7	81
MP3 Player	N/A	N/A	27
DVRs	N/A	N/A	10
Flat Panel TV	N/A	N/A	13
Compact Audio	N/A	22	44

Sources: The New York Times, Appliance Manufacturer's Association, Pew Internet

backyards and more communities ban hanging clothing outside to dry. Each of these factors changes the use of energy and drive consumption.

A similar trend will be the advance of medical devices in residences such as home dialysis machines that are now being encouraged for those that can afford them, as studies are showing that daily dialysis is far more beneficial than the traditional weekly trip to the hospital. People using home dialysis are likely to live longer and be better able to keep their jobs—a boomer trait.

Although the average household size has remained virtually the same, home sizes have increased by 225 square feet and the number of TVs per household has risen from 1.7 to 2.5. We can expect that, as boomers age, they will continue to demand all the space, gadgets, toys and tools that they always have and more. The added load from home healthcare and technologies not even imagined now will only add to this equation. In addition, their children have followed (and, most likely, generations to come will follow) in their footsteps in living an electrically voracious lifestyle.

Load Management, Power Quality and Reliability: The Technology Opportunity

In addition to larger and multiple homes, the continued expansion of the size and number of technologies migrating into the home and the influx of home offices and small offices will continue pressure on an already stressed system.

Demand response will emerge as a key solution to load growth. With an aging population working more at home and using medical devices heretofore limited to hospitals, the desirability and ability of power curtailment changes radically. For all of the sophistication of today's planning models, the projected demand numbers may prove to be wrong. Real-time data, new technology deployment, sufficient futuring and understanding the megatrend of demographics are necessary to successful load management.

While some in the industry are correctly hailing the advance of plug-in hybrid vehicles as a load-balancing solution and revenue generator, the change this will make to the load patterns can be fairly dramatic and unpredictable. Assumptions will be challenged by a generation used to challenging all assumptions. Planning for plug-ins to be used only at night when costs are low and capacity is high may be wrong—especially if gasoline hovers above the $3 mark for long.

There are a number of critical issues for power companies due to this particular brand of boomer load growth. Load demand is just the beginning. The bigger picture reveals an even more important issue of power quality and reliability. The continued expansion of the workplace into the home, of power-sensitive entertainment equipment, and the general growth in new devices means a different opportunity and challenge for power companies and/or to the entrepreneurial marketplace.

As the energy demands for home healthcare rise, so does the need to keep that power on in critical, life-threatening situations. The monitoring of power quality and reliability will require customers to pay the price for such a service. Just as has taken place in the commercial and industrial marketplace, if the incumbent utility is not willing to provide this service, the market will ultimately provide for it in other, entrepreneurial ways.

The opportunities here are not difficult to find but may be difficult to incorporate and manage in a traditionally structured utility environment. New ways of doing business with different types of partners will make the provision of services challenging—but also rewarding financially. For regulated utilities it will mean new agreements with regulators; for public power, recognition of the need to provide advanced and somewhat more risky programs. Some opportunities include power quality services, demand response programs, back-up generation services, energy efficiency programs, load aggregation services and risk management.

Traditional providers will not have all of the answers, but will, at least for the foreseeable future, have the accountability and opportunity to provide the power and power services demanded by the growing retired boomer population.

Succeeding with the Destiny of Demographics

From an aging workforce to population shifting and increased demand, the destiny of demographics will have the most profound effect on the operation of and opportunities in the power marketplace. As the boomer generation continues to change everything in its path, so too will opportunities be created, traditional systems and models challenged and new players encouraged to join the party.

The challenges in the three facets of this megatrend are interconnected: utilities will have fewer workers to manage more complex systems; population will continue to shift to warmer climates with higher energy needs that must be met in as green a manner as possible and the type of demand along with expectations will continue to rise. Responding to these challenges will require innovation in thinking, innovation in technology, and innovation in system operations from planning to new services.

Yet this megatrend offers significant advantage to incumbent players who are established in broad, integrated energy delivery systems and who have an understanding of how the game is played. Technology and vision are at the core of a company's ability to capitalize on these opportunities, especially in those jurisdictions where regulatory recovery and support is still strong. The right investments give the electric suppliers of today a built-in advantage to the boomer generation and those who follow.

Where to go From Here

As we discuss in Chapter 3, with this advantage, power companies should build on their core business in order to strengthen their base through strategic investment that not only expands the company, but reinvents it. The electricity business, like every other, needs a solid base of commodity operations. It also needs to take some risk in reinvention and expansion, particularly in technology, before coming full circle in using that invention as a new foundation from which to build toward the future.

While we have provided examples of specific opportunities which can target the needs outlined in the three areas of this megatrend, strategic planning which incorporates several of these investments within the structure of competitive regulation can lead to significant business growth and protect the market space available to these utilities.

- The implementation of advanced metering systems, load control, distribution automation and a build out of the intelligent infrastructure to achieve digital control and management of the infrastructure lie at the heart of the opportunities to manage through this destiny. Whether it is the replacement of workers with technology or the provision of power quality services to the marketplace, the main focus requires the leveraging of hardware and software investments in new ways. Partnerships will develop (either through regulatory mandate or anticipatory planning) that allow infrastructure to be exercised in a unique fashion. The revolution in telephony has been IP and cellular phones—leveraging infrastructure built for another purpose around traditional business models. The same will hold true for electricity.

- The dearth of engineers will initially force U.S. companies overseas—ironically at a time when the demand for those workers in their native countries is increasing exponentially at comparable pay rates. IOUs will also find some availability in municipal utility engineering staff who take advantage of retiring at 55 and who have at least 10-12 more years in which they desire to work.

- Huge opportunities exist in environmental control installations for regional power plants. Even recognizing there will be a new mix of generation for the future, the demand for support services in this arena clearly outstrips the supply. This will allow utilities with existing expertise to leverage it across other companies—even competitors.

- It is important to recognize that the customers of today and tomorrow will require differing levels of service. By preparing to provide those services on a differentiated basis other than price, companies can position themselves to meet this growing need. In open market states such as Texas and Pennsylvania, the differentiator needs to move away from price (since there is really little difference) and into levels of service and customer engagement. Moving the business and the perception away from the commodity provision of electricity at the lowest possible price means establishing a new "value" on the service provided. This can take various forms (and already has) from offering "green" electricity to PQ services for the home office/small office, market or lifeline services to those with critical medical needs. As technology allows demand response to expand, customers will want to engage much as they do for other services such as cell phones, where "free nights and weekends" shift call volumes.

- In an era of tight supply of capacity and human capital, tight management of the grid becomes even more critical. Interestingly, this also comes at a time where there is a movement to alternative energy solutions which will further strain the resources of the incumbent utility to support the needs of the marketplace.

These investments, properly supported by the current regulatory models, in effect give power companies a free pass into the future and can ameliorate the challenges of an aging workforce, the increases in demand and build stable core business for years to come.

As much as there is a challenge presented from the destiny of demographics, the bigger threat lies within the current utility model which discourages risk, places a premium on the status quo and prefers to plan based on the past (where there is certainty) than in real-time. In the 1990s many urban power companies found themselves in a load-growth mode despite flat or declining occupancies

Destiny of Demographics
Investment Model

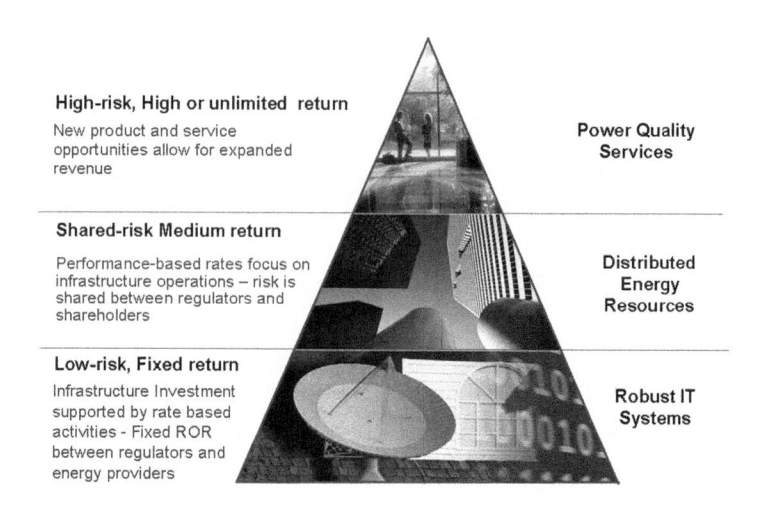

High-risk, High or unlimited return
New product and service opportunities allow for expanded revenue

Power Quality Services

Shared-risk Medium return
Performance-based rates focus on infrastructure operations – risk is shared between regulators and shareholders

Distributed Energy Resources

Low-risk, Fixed return
Infrastructure Investment supported by rate based activities - Fixed ROR between regulators and energy providers

Robust IT Systems

Figure 4-6. Achieving an investment balance is necessary to create infrastructure that can support advanced technologies and services.

and new construction. The reason: the rapid proliferation of computers and related devices. Interestingly, this proved to be even a bigger challenge than usual due to the rate freezes instituted mid-decade.

In 2007/8 the industry has a unique opportunity to expand its asset base to support the demographic megatrend. After decades of underinvestment, there is interest and support for massive amounts of asset increases. More dollars than ever before are chasing new technologies to support the emerging energy landscape. In an era when the value of undepreciated assets is at an all time high, the opportunity is open to investments that will match the onset of advanced communications and intelligence to the next major infrastructure frontier.

This builds on this book's assertion that market trends should be reviewed carefully before investments are made—many will not be successful. Chasing service businesses was in vogue during the early 2000s and many failed due to a lack of worker experience.

Similarly, the early market trend to simple automated meter reading devices, while revolutionary in 1995, can still be justified today based on first cost—but will prove to be an expensive way station on the path to intelligent infrastructure.

The companies that will be successful in using the megatrend of the destiny of demographics will be those believing in the true spirit of business reinvention—that of investing in technology and people, looking beyond the past in planning, understanding the marketplace, visioning opportunity and embracing risk.

References
An Aging Workforce, *EEI Electric*, September/October 2005, by Mark Bridgers and Heather Johnson

Managing an Aging Technical Workforce, *EnergyBiz Magazine*, May/June 2005, Wanda Reder

The U.S. Department of Labor via Utility Jobs Picture Brightens, *EnergyBiz Insider*, Kieth Silverstein, January 19, 2007

Managing the Mature Workforce, The Conference Board Report #1369, September 2005

Michael J. Quillen, president and CEO of Alpha Natural Resources and treasurer of the Virginia Coalfields Economic Development Authority (CEDA), in a briefing to the Commissions Nuclear Energy Institute

Work Force Planning for Public Power Utilities, American Public Power Association Report via Carnegie Mellon University Electricity Industry Center

Issue Alert, The Power of Business Transformation, Jon T. Brock and Matt Smith, January 2006

Black & Veatch, 2006 Utility Industry Survey by Sierra Energy Group

Annual Energy Outlook, Energy Information Administration, U.S. Department of Energy

Aging Work Force Poses Nuclear Power Challenge, Hyun Young Lee, the *Wall Street Journal Online*

Energy Business, Wanda Reder, Managing an Aging Technical Workforce Science and Engineering Indicators, published by the National Science Foundation

2006 Strategic Directions in the Electric Utility Industry, Sierra Energy Group for Black & Veatch

Aging Workforce in the Aging Wholesale Energy Business Abstract, December 2005, Jill Feblowitz, Program Director of Energy Insights

Energy Usage Study, Appliance Manufacturer's Association

Baby Boomers and Real Estate: Today and Tomorrow, National Association of Realtors study by Harris Interactive between, 2006

The New York Times, Ideas & Trends, September 29, 2004

Chapter 5

The Destiny of Carbon Constraints and Capacity Demands

A Megatrend of Conflict

W*ithout prices being set, nature becomes an all-you-can-eat buffet, and I don't know anyone who doesn't overeat at a buffet.*
Richard L. Sandor, Chairman and CEO, Chicago Climate Exchange

In the early years of the 20th century, Samuel Insull aggressively exploited economies of scale to build ever-larger power plants. The result was a revolution in the technology of energy supply, one fully as important as Edison's inventions and the transformer in making technically possible cheap and abundant electricity.

The need for expanding capacity and the movement towards a carbon constrained world are two trains on separate tracks—tracks that are not currently on the same path. While immense effort is needed on both issues, it will be essential to our energy future that these tracks merge onto the same course within the next 5-20 years.

Nearly 35% of the U.S. may experience electric generation capacity shortages by the year 2014—this assuming the rapidly escalating demand curve in electricity is ameliorated by energy efficiency, demand response and the active management of intelligent infrastructure. There has been limited new power plant construction in the last 15 years with the exception of natural gas plants which are susceptible to a volatile fuel and expensive continuing costs. In order to fulfill the country's energy needs in 2030, the industry will need to increase generation capacity by about 30% according to current projections from the Energy Information Administration's (EIA) 2007 Annual Energy Outlook.

The electricity needs of the country, while an immense challenge in itself, must be balanced against the growing call for carbon constraints. The public debate over the need to take action to combat global warming is well over as regulators and legislators are increasingly calling for action. Unfortunately, the significant and almost unmanageable cost impacts of curbing CO_2 are not widely understood. The capital expenditure to build new clean generation, sufficient and efficient transmission as well as carbon capture technologies can be immense.

The good news is that efforts to undertake more serious and far reaching energy efficiency and demand response programs have the potential, if undertaken, to severely lessen demand growth and provide an urgently needed buffer as the industry comes to terms with this impending issue (Figure 5-1).

The electricity industry will likely be required to take the lead in implementing a new carbon free generation fleet which will have huge economic costs even as the benefits are not clear. While there is now a large amount of venture capital and other investment pouring into renewable and clean coal technology, short-term development of carbon free technologies in adequate quantity and efficiency to sufficiently meet expectations is unlikely. And a balanced portfolio will be necessary to ensure adequate reliability and provide for national security.

Renewable portfolio standards are a reality now and there will be more legislation and regulation around carbon emissions. As this develops, integrated resource plans, once a balance between supply and some demand-side resources, will grow in complexity as companies are forced to compete for the same limited set of clean generation resources. The renewables market, which has been spurred by concern over global warming and stimulated by government subsidies, may make the cost of new nuclear seem

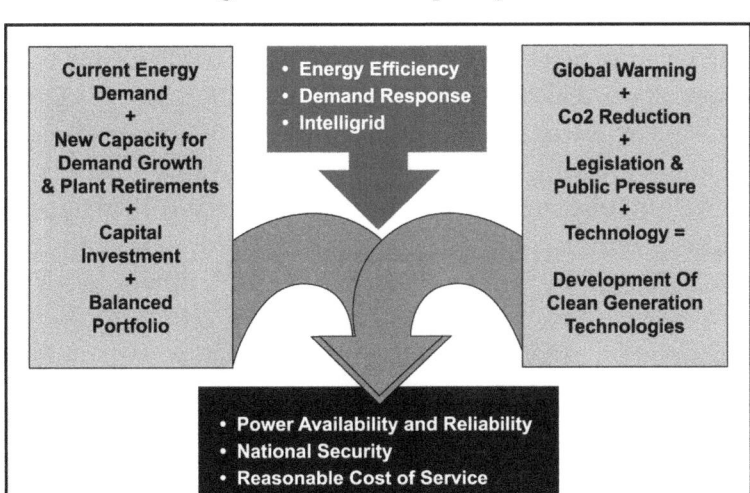

Solving the Carbon/Capacity Conflict

Box 1 (left):
Current Energy Demand
+
New Capacity for Demand Growth & Plant Retirements
+
Capital Investment
+
Balanced Portfolio

Box 2 (top center):
- Energy Efficiency
- Demand Response
- Intelligrid

Box 3 (right):
Global Warming
+
Co2 Reduction
+
Legislation & Public Pressure
+
Technology =

Development Of Clean Generation Technologies

Box 4 (bottom):
- Power Availability and Reliability
- National Security
- Reasonable Cost of Service

Figure 5-1. Solving the carbon/capacity conflict will require efforts on three fronts: 1) capacity additions, particularly clean options, 2) energy efficiency, demand response and smart grid efficiencies and 3) expanded technologies to reduce carbon and develop new clean and renewable generation technologies. Only through this will the energy enterprise will be able to provide adequate power availability, reliability, security and quality, a reasonable cost of service and appropriate national security.

inexpensive by comparison. As can be evidenced in methanol development efforts, economic logic is not always the priority in the face of global warming. The growth in hybrids, this in contrivance to financial logic, shows the market is being pushed inexorably in this direction due to environmental concern regardless of economic reality.

It is very likely that there will be an economic tool used in the "fight" against carbon, whether it is a trading scheme or a tax, the impact will be very real on the price of electricity. Trading exists today in a world of offsets—regardless of whether the reality of those offsets meets the test. As some would point out, "locking" carbon in forests is only a temporary solution as the ultimate death and decay of those assets will eventually release carbon anyway.

In order to meet the electricity demands of the future, we must consider all sources of energy. Even though the price shocks of natural gas have faded, the potential for volatility is there. Even as investment is growing in the renewables market and there is talk of new nuclear construction, the "inconvenient truth" is that coal will continue to remain a mainstay of generation for at least the next 20 years. Although an honorable goal, the future power needs of the U.S. cannot currently be fulfilled in the foreseeable 10- to 25-year time frame by renewable resources, demand response and energy efficiency alone, no matter how robust.

Putting this all together, we can see the obvious challenge:

- necessary new capacity to serve increased electricity demand and to replace aging generation assets;

- the need for a balanced portfolio to ensure affordable energy, reliability and national security;

- concern over global warming pushing CO_2 reduction efforts and green technology development, and

- the capital-intense financial requirements in order to fulfill all of this.

This chapter looks at the megatrend of the carbon and capacity conflict in progress in which the growing need for electricity conflicts with increased concern over what to do about global warming. While there are many approaches to partially solve this issue through demand-side management, technology and legislation, there remains a huge question—How will we generate the power necessary to meet our needs while reducing carbon output? Solutions to this dilemma provide many opportunities to the savvy, forward-thinking power company or entrepreneur, both through technology investments and new ways of doing business.

Supply-side Issues and the Carbon Conflict

Total electricity demand in the U.S. is projected to grow by 1.5 percent per year from 2005 to 2030 to 5,478 billion kilowatt hours. This represents a 1.7 billion kilowatt-hour increase from 2005 and a major leap in the consumption of electricity (Figure 5-2). To meet this demand growth and replace older, inefficient fossil plants, the country will require an additional 292 gigawatts of generation capacity. This is equal to roughly 365,800 MWh coal fired power plants costing $1-$1.5 billion each—and the cost of new construction is rising. Between now and 2015 the EIA estimates the U.S. will need an additional 50,000 MW of new generation capacity. Given the lead time for permitting, the challenge of siting and the hesitancy in new plant construction, the chance

Electricity generation by fuel, 1980-2030 (billion kilowatt hours)

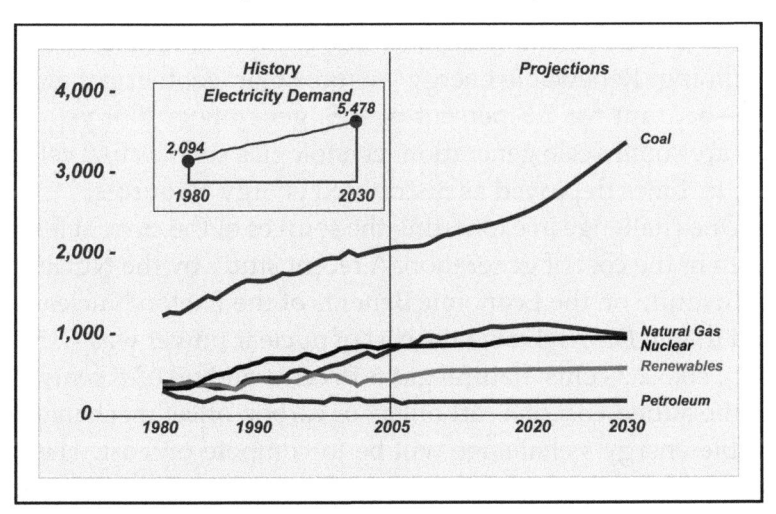

Figure 5-2. Electricity demand is expected to increase in the U.S. by 1.7% per year through 2030 requiring an increase in generating capacity of roughly 30%. *Source: Energy Information Administration 2007 Annual Energy Outlook*

of this being met with anything other than additional combined cycle natural gas plants (plus efficiency) is slim—and continues to put the nation at risk.

Expanding the portfolio of generation sources will be critical in order to assure a diversified portfolio, the greatest security and least cost of electricity supply. Providing an infrastructure that allows for the wide deployment of distributed power sources including renewable energy sources as an integral part of the electricity supply system will be important in reaching these goals. And, as stated, there is a continued need for coal and nuclear, even as the coal-fired generating fleet is past its prime and in need of replacement and concerns of global warming mount.

The current U.S. generation mix is dominated by large, central station power plants. Coal plants provide nearly half of the electric power consumed in the United States, while nuclear power accounts for one-fifth of the mix. Natural gas plants generate about 19 percent of the nation's power, and this share is projected to increase as gas-fired combined cycle turbines dominate new capacity additions. Hydropower provides about 6.5 percent of the nation's electricity. Oil-fired plants contribute about 3 percent, and this share is declining. Renewable energy—wind, solar, geothermal and biomass—account for 2.3 percent of U.S. generation. Not yet on the radar are small-scale generation technologies such as fuel cells and micro turbines deployed as distributed energy resources.

One challenge in expanding the sources of the current fleet can be seen in the cost of generation. A recent study by the Nuclear Energy Institute on the Economic Benefits of the Exelon Nuclear Fleet found that in Pennsylvania the cost of nuclear power was 1.15 cents a kWh, coal 2.7 cents, natural gas 6.18 cents and oil 6.73 cents. Even with the added cost of a carbon tax or carbon offset mechanism, renewable energy's challenge will be to compete on cost. The issue is further exacerbated by the reliability of renewables. During the 10-day "heat storm" in California in 2006, wind was unavailable as well due to climatic conditions.

There are between 30-37 nuclear power plants that are working through the process of permitting on their way to construction.

NRG Energy formally announced its first plant in September, 2007, in Texas, near the site of its existing South Texas project. Interestingly for the industry, NRG has never constructed a nuclear power plant. It is estimated that the investment in new nuclear units will range from $60-$90 billion over the next 15 years.

Uncertainties about the direction of future industry regulation and energy/environmental policy have made long-term planning and investment decisions for utilities extremely difficult. The onset of industry deregulation triggered a boom in the construction of new generation capacity. The exuberant overbuilding flooded the market with available supply, lowering the price of electricity and leading to a wave of cancellations in new plant construction projects. Such boom-bust cycles contribute to price volatility, reduced reliability and energy security. Furthermore, they underscore the consequences of a focus on short-term profits in the absence of a regulatory framework that allows for long-term planning.

The most significant long-term issue affecting electricity supply is CO_2 emissions from fossil-fired plants. The structure and timing of any CO_2 policy will determine which generation technologies will be economically preferred in the long run. In the absence of clear direction on CO_2 emissions management, companies are understandably hesitant to commit to the construction of large generating plants.

Industry deregulation and air-quality regulations have favored the installation of gas-fired combustion turbines that offer low capital cost and roughly 1/2 the CO_2 emissions of a conventional coal plant. As more gas-fired units have come on-line, the nation's existing coal-fired plants, originally designed for base load operation, are being operated as cycling or peaking units. This cycling duty increases wear and tear on the aging fleet of coal units and also makes their operation less profitable. Industry experts predict that many of these plants will need to be retired sooner than planned, further tipping the gas/coal balance in favor of gas.

Despite the nation's growing reliance on natural gas for power generation, especially in new combustion turbine power plants, gas is not going to eliminate the need for coal. Projections from EIA show that coal will continue to provide about 50% of power generation in the U.S. for the foreseeable future, as well as remain the backbone of power generation in other nations with extensive coal resources, notably China (Figure 5-2).

The boom in gas turbine construction and the prospect of accelerated coal plant retirements have led to the assumptions that all new plants will be gas-fired, that no new nuclear plants will be built, and that coal is a fuel of the past. But these assumptions are misguided. In fact, the boom in gas turbine construction has led to price increases and supply shortages, underscoring the need for fuel diversity to ensure national energy and financial security—just as portfolio diversification helps mitigate the risk of investing in the stock market.

Another key issue influencing this megatrend and energy supply decisions involves constraints on the transmission and distribution system, described in Chapter 6, The Destiny of Intelligent Infrastructure. High costs, public opposition, and regulatory uncertainties have blocked the construction of new transmission lines, including those that might connect new renewable assets to the grid. This situation, along with load growth accompanying population expansion, has contributed to transmission congestion and bottlenecks that impose further limits on the siting and construction of new generation.

The Carbon Conflict: Unclear Policy, Uncertain Directions

Without carbon reduction technology, CO_2 emissions from electricity generation are projected to increase by more than 40% between 2005 and 2030 to more than 3,300 million metric tons. This is due primarily to the expected increase in coal power production. At this time, as many as 130 new coal plants are being proposed across the U.S., and more will be needed if we are to meet the growing demand for electricity.

In early 2007, TXU's proposed buyers announced a decision to back out of building 8 of 11 proposed pulverized coal units with advanced environmental controls. While this recent event has cast doubt on the number of greenfield pulverized coal plants that will actually be built, there will be a growth in units nevertheless as demand outstrips current supply. With IGCC technology close to prime time, companies are making hard decisions on balancing their more immediate need for capacity and the environmental impact. Interestingly, any lack of new construction will mean that older, inefficient plants will continue to run and may increase in value as the supply dwindles. This can have an unwanted effect on carbon numbers as mothballed units come back into the market. The next five years will be a challenging time for strategic planning in generation, while there is a transition from reliable, traditional coal technologies to advanced IGCC and carbon sequestration technologies still being refined to commercial level and construction capability.

The sense of urgency, the repercussions on conventional coal plants from the TXU decision and the expectations regarding carbon capture have changed dramatically. Initially there was talk of having IGCC plants "carbon capture ready." Now the assumption is clearly that IGCC _must_ have carbon capture in order to move forward.

Currently, the issue of carbon capture has been driven to the top of the political agenda with bi-partisan support from the industry and environmental groups coming together around the realization that capacity needs cannot be met with renewables and conservation alone. There has been quite a bit of discussion—and almost universal agreement—that the days of conventional PC plants not already in the permitting process were over. One particularly interesting market driver is emerging: the defensive need to protect carbon exposed assets by supporting IGCC and carbon capture sequestration (CCS). This risk mitigation strategy is being deployed (or at last examined) by companies whose holdings include power and chemical plants, mines and other carbon generating facilities. The strategy is to protect portfolio value by investing in IGCC with CCS thereby picking up offsets.

Carbon Prices

There are numerous predictions of carbon tax and/or cap-and-trade affects which have estimated the cost of carbon as from the current industry assumption of $7-$20 a ton to north of $100 a ton by 2030 and even to $800 a ton by 2100. An analysis of current bills in Congress shows costs ranging from $7 to $39 by 2015. Recent modeling by R.W. Beck has found that it will take a $50 per ton credit/tax to really effect change in the market—and this certainly has huge impacts on the economy, especially if energy is singled out. All of the models being reported make some bold assumptions that include significant advances in technology, drivers from a policy side that include economic inducements to motivate otherwise uneconomic technologies and massive infrastructure additions to handle carbon off take. While ballyhooed as being "clean," natural gas has its own carbon problems. With rising carbon prices, natural gas may become uneconomic as well. The debate continues on carbon prices. It appears that 90% capture may not be enough, even assuming technology catches up with desires. Without CCS, some posit coal could be completely gone as a fuel source in the U.S. by 2040-2045.

Managing the capacity-carbon challenge will be the preeminent environmental issue facing the electric power industry. Although the future impacts and timing of climate change remain uncertain, reducing emissions of carbon dioxide and other greenhouse gases will almost certainly be essential to a sustainable energy future. It is likely that limits on CO_2 emissions will become mandatory. Climate policy will affect a company's technology choices, investment decisions, and compliance strategies. On a broader scale, our response to the climate change issue will have major implications for the nation's fuel diversity, energy security, electricity costs, and global economic competitiveness. Given both the uncertainty and consequences associated with climate policy, companies face near-term challenges in managing risk by understanding the issues involved, keeping abreast of developments and preparing for a range of possible control scenarios. Over the longer term, companies will need to develop and deploy advanced, low carbon technologies and/or adopt other CO_2 reduction strategies.

The climate change policies that will be established over the coming decades on the international, national, state and local levels will be one of the key determinants of how global climate change is addressed. The Bush Administration rejected a cap-and-trade policy for CO_2 by announcing in March 2001 that the U.S. would not ratify the Kyoto Protocol. In its place, the Administration's climate policy, which was announced in February 2002, provides incentives for voluntary actions that reduce greenhouse gas emissions and increase carbon sequestration. The policy sets a goal of an 18% reduction in GHG intensity (GHG emissions per unit of economic output) by 2012. A number of bills setting mandatory GHG emissions levels for future years have been introduced in Congress. Many U.S. states, regions, and municipalities have established, or are considering, various policies to address climate change. California, Florida and states in the Northeast are working to impose their own restrictions on carbon, which further adds uncertainty to an industry known for being shy in the face of risk.

In addition, a number of U.S. companies, including electric utilities, have actively participated in voluntary reductions of CO_2 gases. In the year 2000, these efforts resulted in reductions of 237 million metric tons of greenhouse gas emissions, or roughly 10% of the projected 2 billion metric tons of CO_2 in 2005. This exceeded the aggregate pledges in that year by 36%.

But more than voluntary reductions will be needed. About 80% of the electricity generated in the U.S. is derived from hydrocarbon fuels whose combustion releases CO_2 into the atmosphere. Technology investment will play a key role in finding a solution to how to curb carbon emissions. The Electric Power Research Institute (EPRI) estimates that CO_2 recovery would add 60-70% to the cost of electricity for conventional gas or coal-fired power plants, while cutting plant efficiency by up to one-third. To improve the outlook, work is going forward on several options for these existing plants, including solvent-based scrubbers and advanced membrane separation. Advanced clean coal technologies, however, are showing much promise. Research indicates that in new integrated gasification combined cycle (IGCC) plants, combined with CO_2 capture through geologic sequestration would increase electricity costs by only 25%—suggesting that IGCC would be competitive with gas-fired combined cycle plants (with CO_2 removal) when gas prices are above $4/MBtu.

However, significant investment will need to be made now in order to meet reduction needs within the next 20 years.

Overall, stabilizing atmospheric CO_2 concentrations will require a three-pronged approach (Figure 5-3) involving policy, economics, and technology. Policy will answer questions about when, where, how, and who. Economics will influence the selection of options, including emission-trading credits that offer the greatest benefit per dollar spent. Technology investment will refine CO_2 capture and confinement, develop carbon-free generating options, and increase end-use efficiency.

Fueling the Future: Facing the Conflict

Economics, environmental impacts, safety, availability, and public perception are among the key competing factors in evaluating fuel sources for power generation. There is no perfect fuel,

Controlling CO_2 Emissions: A Three-pronged Approach

Stabilization of Atmospheric CO_2 Concentrations

Policy
When, Where, How, Who

Economics
Emission-trading credits

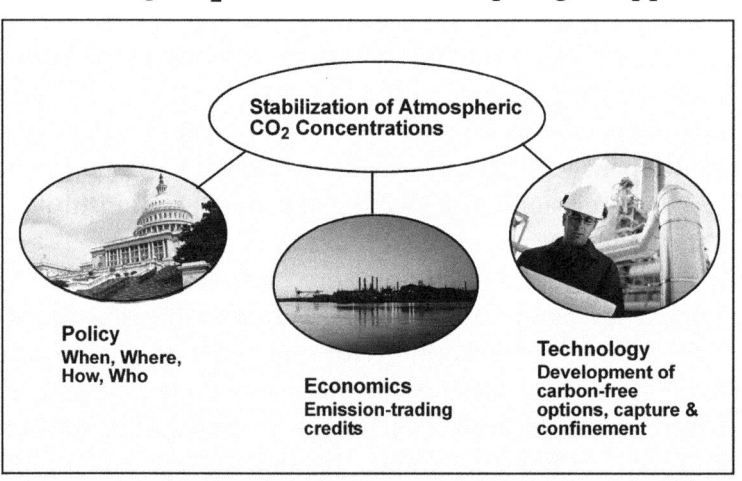

Technology
Development of carbon-free options, capture & confinement

Figure 5–3. In order to stabilize CO_2 atmospheric concentrations, policy issues will need to be addressed. In addition, the process will be influenced by economic impacts and technology investment. *Source: EPRI*

just as there is no perfect technology for power generation. Coal is abundant and inexpensive, for example, but its combustion releases pollutant emissions. Nuclear doesn't have coal's emissions, but suffers from public concerns over plant safety and uncertainties over waste disposal. Natural gas is clean and efficient, but increased demand raises questions about availability and price volatility. Even pollution-free green renewables have problems such as land-area requirements of solar facilities, deaths of birds, including endangered and protected species, from wind turbine blades and technical issues with intermittent power sources.

Fortunately, technological enhancements and new opportunities are available or emerging that can help mitigate the problems with each fuel source so companies can leverage the strengths and minimize the weaknesses of each. Given the need for a balance, it is best to assume a fuel-neutral perspective, focusing on the most effective strategies for business, society and national security. Viewing fuel balance as critical allows the positioning that it is the best interest of our nation's energy and financial security, and the competitiveness of the energy sector, to optimize fuel diversity.

Electricity generation in the U.S. relies on an array of primary fuel sources: coal, uranium, natural gas, water, oil, and renewable energy sources, including solar, wind, geothermal, and biomass (Figure 5-4). A diverse portfolio of fuels—like a diverse portfolio of investments—helps reduce risk exposure and provides flexibility for meeting future uncertainties. The price and availability of fuels, technological advances, and changes in regulations or policy fall among those future unknowns.

The U.S. electric power industry burns about $30 billion worth of fossil fuels annually. As competition heats up, there will be increased volatility in fuel prices. Like a sound investment portfolio, optimizing a strategic, diversified fuel mix is critical to the profitability of power plants, to maintaining stable electricity prices, and to assuring our nation's energy and financial security.

It will take a combination of many approaches for the electricity sector to contribute to CO_2 reduction which will be discussed as solutions to this megatrend of carbon and capacity conflict. The

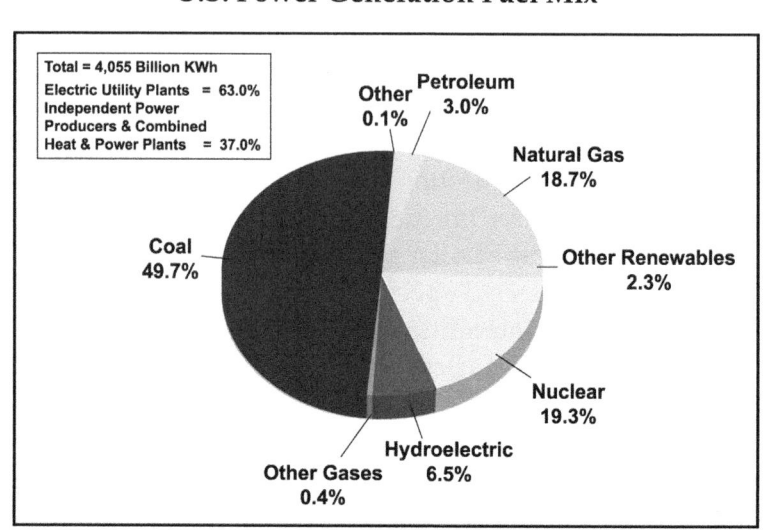

U.S. Power Generation Fuel Mix

Total = 4,055 Billion KWh
Electric Utility Plants = 63.0%
Independent Power Producers & Combined Heat & Power Plants = 37.0%

- Other 0.1%
- Petroleum 3.0%
- Natural Gas 18.7%
- Other Renewables 2.3%
- Nuclear 19.3%
- Hydroelectric 6.5%
- Other Gases 0.4%
- Coal 49.7%

Figure 5-4. Fuel diversity is critically important to the nation's economic health and its national security. Factors used in determining fuel use range from safety, availability and public perception, yet energy policy is needed to balance the fuel mix to meet national energy and economic objectives. *Source: U.S. Department of Energy, Energy Information Administration.*

comparison chart which follows (Table 5-1) shows some of the fundamental requirements, output and environmental impact of different generation options.

Together, these represent a portfolio of options—each with its own advantages and disadvantages from an economic, environmental, and social perspective. Therefore, different stakeholders hold a broad spectrum of views on which options should be pursued, and on what types of public policies should be implemented to encourage or discourage various options. Some of the major issues integral to climate change are discussed below.

In the end, the nation needs a reliable supply of reasonably priced electricity generated with the cleanest technology possible to minimize adverse environmental impacts and contribute to CO_2 reduction. No single form of generation can meet all the nation's

Table 5-1. Chart comparing different generation technologies—cost, footprint, and limitations (such as region, social acceptance or intermittence). If possible, include residential solar and demand response programs in this chart.

Power Generation Technology Comparison

Plant Type	Capacity MWh	CO_2 Ton/MWh	Footprint MW/Acre	Reliability	Location Dependent	Initial Investment $kWh
Pulverized Coal	800-900	1.0	.9	H	No	$2,500-$3,000
IGCC	600	.8-.9	.6	?	No	$2,700-$3,400
Natural Gas	500-600	.4-.6	5	H	Yes	$880-$1,000
Nuclear	1,200	0	.4	M-H	No	$4,500-$5,500
Large Wind	150	0	.2	?	Yes	$1,200-1,800
Solar	500*	0	1.1*	?	Yes	$4,000-$6,000 ??

*Based on planned SCE 500MW plant using Stirling Energy Systems—largest solar installation

needs, now or in the future. A diverse generation portfolio is needed that delivers maximum benefit to both industry, society and the environment.

Another way to look at the situation is to understand the sheer number of plants that will be necessary to offer the benefits of a single IGCC facility—not to mention the acreage required (Table 5-2). An IGCC plant, while an expensive and time-consuming investment, provides a continuity of power and a capacity factor advantage over alternative technologies.

Table 5-2. It would take 850 2 MW wind power plants and 250% the land space to replace one 600MW IGCC power plant.

Generations Sources as Compared to a Single IGCC Plant				
Source	# of Projects	Size	Capacity Factor	Acreage
IGCC	1	600 MW	95%	40
Biomass	30	20 MW	85%	300
Landfill Gas	268	2 MW	95%	400
Wind	850	2 MW	30%	10,000

Source: RW Beck, Inc.

King Coal

Abundant, inexpensive, and indigenous, coal has long been the backbone of U.S. power generation. Coal is the country's most abundant fuel source, and "King Coal" was the primary fuel behind the country's electrification in the 20th century. Coal accounts for more than half the electric power generated in the U.S., and the nation's coal supply is projected to last 250 years. Available U.S. coal reserves are nearly a third of the world total reserves making this power source a stable commodity with relatively little volatility.

The problem with coal has long been its environmental impact. Burning coal releases sulfur dioxide (SO_2) and nitrogen oxides (NO_x), which contribute to acid rain and smog, as well as particulate matter, mercury and of course carbon dioxide. Yet remarkable technological advances in emissions control systems have helped clean up coal's image as a dirty fuel. Emissions of SO_2 and NO_x are down 40% and 20%, respectively, since 1980, even though electricity production climbed by 35% in that period.

The key to coal's long-term viability as a fuel for power generation and a solution to the challenges of this megatrend lies in finding ways to reduce or even eliminate coal's environmental impact of climate change through CO_2 emissions. There are options. Clean coal technologies can reduce emissions and improve generating efficiencies. IGCC, mentioned in the previous section, is a key enabling technology for future coal-based power generation that essentially involves refining coal to produce a clean-burning gas. Combined with carbon sequestration, this option can help neutralize the impact of CO_2 emissions from coal, in two broad approaches: Indirect sequestration involves the biological removal of CO_2 from the ambient atmosphere, for example by planting trees. Direct sequestration involves the separation and capture of CO_2 and disposal in deep saline aquifers. Carbon management may also involve reducing net CO_2 emissions through carbon trading markets.

The Opportunity
IGCC: The Environment-Friendly Coal Plant
As a critical pathway to the future, IGCC plants could enable energy suppliers to tackle regulatory and fuel issues head-on, while staying within their core business and expanding into new markets. These plants gasify coal to produce a hydrogen-rich fuel for combined cycle turbines that generate electricity with very high efficiency (Figure 5-5). IGCC technology offers solutions to the most pressing environmental issues facing power generation. IGCC plants produce low emissions not only of CO_2, but also of NO_x, SO_2 and other pollutants associated with conventional coal plants. Moreover, some advanced IGCC designs now incorporate hydrogen production capabilities. The hydrogen can be sold for use as transportation fuel or to power fuel cells or other small-scale or distributed generators. Some IGCC concepts also allow production of other chemicals such as methanol, which can be used as fuel or as a feedstock for higher-value hydrocarbon products for sale to outside markets.

Depending on the design, an IGCC plant, coal may be first treated with oxygen and steam to create a clean-burning synthesis gas, which is then stripped of sulfur and other impurities. This gas is burned to drive a combustion turbine, which in turn provides hot exhaust to generate the steam needed to drive a steam turbine. Both turbines produce electricity in a combined cycle. These plants are proving to be remarkably environmentally benign—meeting very stringent standards for emission of SO_2, NO_x, particulate matter, mercury and other air pollutants.

Coal Fired Integrated Gasification Combined Cycle Plant

Figure 5-5. Commercial-sized, coal-based integrated gasification combined-cycle (IGCC) power plants are in operation and producing ultra-clean electricity in Tampa, Florida; Wabash, Indiana; and Buggenum, Belgium.

In more recent designs, such as one developed by Southern Company and KBR (Figure 5-6), an air blown system is used in a manner similar to gasoline "cracking" with the resultant gases being used to fire generators. These systems are claimed to have higher efficiencies and lower emissions.

And that's just the beginning. Advanced IGCC designs that incorporate CO_2 removal and hydrogen production could allow for *powerplexes* that produce electricity for the grid, hydrogen fuel for distributed generation and transportation, and salable chemicals. IGCC is thus a transformative technology that can help power producers become integrated energy companies—expanding into new markets while building a bridge to a low-carbon, hydrogen-based energy economy.

The IGCC Air Blown System

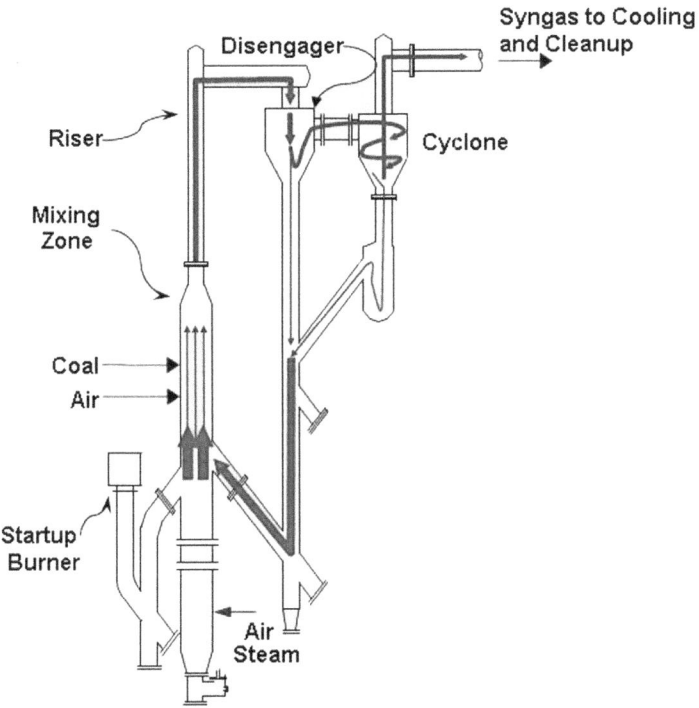

Figure 5-6. A new IGCC technology developed by Southern Company and KBR provides lower up-front capital as well as O&M costs and higher availability than existing IGCC technology. *Source: Southern Company*

Currently there are only four IGCC power plants operating in the world, two in the U.S. and two in Europe. However, although the technology does not yet have commercial status, it is slowly taking hold. A 540MW unit is currently under construction by Lima Energy in Ohio. In addition, nine plants in the U.S. pending approval with 14 more proposed or undergoing feasibility studies.

The U.S. DOE led FutureGen Alliance is in the planning stages on a zero-emission IGCC demonstration facility. Construction is scheduled to begin in late 2008 and the facility should be up and running by 2012. The goal is to reach at least 60-80% efficiency (traditional pulverized coal plants are typically 35-40% efficient) and integrate carbon sequestration technology to bury the CO_2 emissions while also producing hydrogen in the process. The $billion effort, which has been on the drawing board for several years, is run by a coalition in partnership with DOE representing a number of large coal companies and electric utilities including American Electric Power, Southern Company and Peabody Energy. Several foreign governments and the largest Chinese energy company, The China Huaneng Group, are also partnering in the effort.

In a separate project, Duke Energy is working with General Electric and Bechtel Corp. to take an existing 160-megawatt coal plant and convert it into a modern 600-megawatt coal gasification facility in Indiana. The plant will take several years to complete. The benefit of this is that companies can replace older plants that are less efficient and have high CO_2 emissions with more efficient and cleaner technology.

Capturing Carbon

There are interesting assumptions being made concerning "clean coal." While IGCC is the answer for the future, the belief that carbon sequestration is here or near, may not be justified. A number of states are requiring all new coal plants to be IGCC—and that is a logical choice given the state of the IGCC plants today—within the limitation of the industry's capacity to build these plants. What is not clear is how the carbon is to be captured and stored.

Some of the potential approaches to managing carbon have been addressed in this and other chapters. These include improving the efficiency of fossil-fired generation, improving the efficiency of end-use equipment and appliances, and a shift over the long term to an energy economy based low-carbon sources such as nuclear and renewables. Although these approaches will help, they won't be enough. Given that coal produces half the world's electricity, even all the above approaches combined won't be enough to cap CO_2 levels over the next half century. This means we have to look to methods for capturing and sequestering carbon.

The new 110th Congress is expected to take action on greenhouse gas issues in climate legislation and companies like ConocoPhillips, General Electric and Shell Corp. are spending billions to develop not just coal gasification technologies but also the tools to bury CO_2. Carbon can be captured and sequestered by both direct and indirect means. Direct technologies involve capturing CO_2 from energy facilities and transporting it to a disposal site. Indirect means involve removing CO_2 from the ambient atmosphere by biological processes.

Direct capture technology can take several forms, including pre-combustion approaches that remove carbon dioxide from the syngas from oxygen-fired gasifiers, and post-combustion capture that involves scrubbing CO_2 from flue gas.

Several possible options for carbon sequestration are being investigated, including

- CO_2 sequestration in geological formations including oil and gas reservoirs

- Ocean injection and stimulation of phytoplankton

- Uptake by terrestrial ecosystems

- Generation of calcium carbonate by marine animals

- Biomimetic catalytic processes

Carbon Sequestration

The most promising answer to capturing carbon before it hits the atmosphere is carbon sequestration into geologic reservoirs. The largest cost of this approach comes from the process used to separate the gas from the raw coal or, after burning, from its emissions, and then to compress the gas into a liquid form. Some calculations show that carbon sequestration would increase the cost of electricity from coal-fired power plants by as much as 80 percent. This impact could dramatically alter the economics of electric power, particularly in states where up to half the electricity comes from coal-fired plants.

A report sponsored by the U.S. Department of Energy on behalf of several Southeastern energy companies shows that it could cost $4 billion annually to eliminate the carbon dioxide generated by power plants in the Carolinas. The immense cost for cleaning the CO_2 from these coal plants would be equivalent to building two nuclear power plants every year. The problem lies in the fact that this region lacks the proper geology to trap the gas and would require hundreds of miles of pipelines to deliver it to Kentucky, West Virginia or offshore—at a cost of $4 billion.

Yet, where the geology is right, carbon sequestration can be a very convenient and effective tool for indefinite storage of CO_2 emissions. American Electric Power is in this position. AEP announced plans to begin capture of up to 100,000 tons of carbon dioxide a year at a plant in West Virginia and to store it on site in a deep saline reservoir using a lower-cost chilled ammonia application. This is the first of a 3-phase project sponsored by DOE through NETL. Once the technology is validated, it will be scaled to capture 1.5 million tons of CO_2 a year at a 450-megawatt plant in Oklahoma that will link safe, long-term sequestration with enhanced oil recovery by 2011. The technology has the potential to capture up to 90 percent of a plant's carbon dioxide.

In a parallel effort, the CEO of TransAlta, Canada's largest utility and others are building a pipeline to transport CO_2 from oil sand deposits in Alberta for sequestration. By deploying carbon sequestration on a wide scale, the country aims to demonstrate that this technology is effective in both cost and environmental terms.

There are other technologies on the horizon for the capture of CO_2. Cambridge, Massachusetts-based GreenFuel Technologies Corp is working with NRG Energy and Arizona Public Service on feasibility studies and pilot projects that would use specialized algae biomass to recycle the CO_2 back into the plant as fuel, or potentially be processed into bio-diesel or ethanol. Other companies are also working on similar projects that create a near-zero emission plant and also provide a reuse for the carbon by-products.

Return on Investment

The advantages are clear. The low emissions and high efficiency of IGCC, together with the ability to remove CO_2—at markedly lower cost than from traditional pulverized coal or fluidized bed combustion plants—makes IGCC a key enabling technology for future coal-powered generation. Moreover, IGCC exemplifies the "systems approach" to technology deployment as described in preceding chapters. IGCC not only enables a company to generate clean electricity using inexpensive, indigenous and abundant coal, but also to gain NO_x, SO_2 and CO_2 credits for profitable emissions trading. In addition, the ability to co-produce hydrogen could be exploited to power on-site fuel cells (or other small generators) to further boost the overall efficiency of the integrated plant. Excess hydrogen, or other chemicals produced in the IGCC process, could be sold for use elsewhere. The IGCC systems-based approach thus helps a company maximize the value added to its invested capital and purchased fuel.

Companies have been reluctant to invest in IGCC because initial investment in the technology has been perceived to be more expensive on a $/kW basis than traditional pulverized coal plants. However, this has changed over the last several years. New IGCC designs have brought costs down to a level even with pulverized coal plants. The overall cost of building any type of plant has gone up. Both traditional and IGCC coal plants continue to be less cost-competitive with current combined cycle gas turbines on a construction basis. The perception that IGCC costs remain high and the lingering reluctance to implement new technology is understandable, but it's also reminiscent of the industry's reluctance to invest in gas turbine technology in the late 1970s and 1980s when nuclear and coal power were dominant and natural gas was thought to be in short supply.

Circumstances change, and a progressive company will look beyond today's shortsighted focus on natural gas and make the bold moves necessary to get ahead of the game. A key factor is can we build enough IGCC plants soon enough to meet capacity needs. Investing in IGCC certainly carries risks, but it also provides a path that simultaneously addresses looming fuel and environmental issues while unlocking new business opportunities in the 21st century energy industry.

Natural Gas

Natural gas is expected to be the fastest growing component of world primary energy consumption through 2015, according to the EIA's 2006 International Energy Outlook. This generation source now accounts for 19% of U.S. power generation capacity, and seven states obtain more than a third of their power from gas. In recent years, natural gas-fired combustion turbines and combined-cycle turbines have dominated new generating capacity additions—their popularity driven by their low emissions, high efficiencies, low capital costs, and short construction lead-times. Electric generation capacity by natural gas in the U.S. is projected to increase to 22% by 2015, however, as fuel prices rise, more coal-fired plants will be built and this number is expected to recede to 16%. However, as history has shown us, the price and penetration of natural gas continues to confound even the most expert observers. During the 1990s low natural gas prices resulted in a spate of new plants and even a temporary market in combined cycle plant "futures" where companies sold their rights to the next turbine off the line to others needing the hardware. That situation rapidly reversed as prices rose over the $10 mark with machines sitting in warehouses looking for a home.

It is important to note that natural gas competes for other uses besides power generation. Natural gas is used as an industrial fuel, as a chemical feedstock, vehicle fuel, and for residential and commercial space heating and cooking. The surging demand by all sectors is driving up prices and highlighting the most significant long-term challenge relating to natural gas—whether adequate supplies can be provided to meet sharply increased projected demand at reasonable prices. Natural gas supply constraints and price increases could adversely affect power reliability as well as limit the market for emerging distributed energy resources such as fuel cells and micro turbines.

The prospect of gas shortages and price increases underscores the importance of a diverse generation portfolio. Gas price volatility may support the continued viability of coal and nuclear power plants and should stimulate research and development of advanced, energy-efficient supply options.

Gas lives in an interesting unbalanced market. When prices are low there is little incentive to explore new sources, extract more expensive reserves and transport it to market. This is the exact time when electricity producers are most interested in natural gas as a fuel source. This is coupled with an increase in the use of natural gas in industrial processes as well as in heating. As this demand forces prices upward, the attractiveness of gas as a fuel source wanes. Industrial customers complain of high fuel costs driving up their production costs; homeowners fume over the rise in their prices and electricity suppliers in regulated markets are unable (at least for a proscribed period of time) to pass along those increased costs to consumers. Further, the gas price increases make coal and nuclear even more attractive.

Natural gas is just a part of the carbon/capacity megatrend. It has enjoyed an era of temporary popularity. With shorter lead times for construction approvals and lower investment commitments, it has been a logical choice for many utilities and power producers. However, as prices go up, it will become a less attractive choice. The availability of natural gas in the short term makes it popular when prices are low but, as many utilities found in 2004/5, prices move in both directions.

There is significant hope for liquefied natural gas (LNG) to relieve some of the price pressure. However, LNG has its own issues, not the least of which are the locations where it is found—in countries with political uncertainty and an anti-U.S. sentiment. This is coupled with siting difficulties for regasification, and, even while the carbon output of gas plants is lower than that of coal, the reality is that they are much higher than the common perception.

Flocking to natural gas when it was a low construction-cost option was a market trend in response to the carbon/capacity megatrend. Utilities saw this as a way to possibly circumvent the coming conflict. Although, in reality, natural gas produces 40%+ of the carbon that a pulverized coal plant does—not nearly the emissions, but also not as clean as some might have liked to believe. Using more natural gas does not take advantage of the megatrend—nor does it eliminate the challenge presented by the carbon/capacity conflict.

Nuclear

Nuclear energy accounts for 20% of U.S. electricity generation. The last nuclear generating unit was brought on line in the U.S. in 1996 and although the number of U.S. nuclear plants has declined since then, nuclear generation has increased steadily in recent years. There has been fluctuation in generation output resulting from up-rating of existing units, retirements and fuel prices, however, nuclear posted a record year in 2004 of 788.5 billion kilowatt hours, 1 % more than the last record set in 2002 and is expected to reach 795 billion kilowatt hours in 2007. This represents continued growth in an industry that was producing less than 600 billion kilowatt hours per year prior to 1991.

Global warming has created renewed interest in the nuclear generation option. Add to this, rising natural gas prices and incentives from the 2005 Energy Policy Act, and that interest has grown. As of the end of 2006, a number of companies and consortiums were seeking permits and the necessary approvals for at least 30 nuclear plants. These will take time, however and ultimately, only a handful of these will likely be constructed. EIA estimates that 9 gigawatts of nuclear capacity will be added by 2020 to qualify for EPACT incentives and another 3.5 gigawatts will be built by 2030. Meeting EPACT deadlines will be no easy task. Additionally, companies face a reluctant investment community without more assurance of cost recovery from state regulators. The primary uncertainty for low-cost financing is that after several decades, the cost of a nuclear plant is essentially unknown.

Another roadblock for nuclear power was touched upon in much more depth in the Chapter 4, The Destiny of Demographics—that of a trained workforce. With half the nuclear workforce more than 47 years old and facing retirement, and even key suppliers to the industry—architects, engineering firms and fuel suppliers on a similar tract, the industry has a serious workforce hurdle to overcome. Nuclear energy also faces long-standing challenges, particularly public perceptions of plant safety, and the long-term disposal and storage of spent fuel and radioactive waste. The March 1979 accident at Three Mile Island is one of the primary impediments to a nuclear renaissance.

However, as a strategic approach to this megatrend, the promise of nuclear power as a generation source free of the emissions associated with fossil fuel combustion, including NO_x, SO_2, mercury and CO_2 is a solid one. The cost of electricity generated by nuclear plants is cost-competitive with that produced by other sources. In fact, with capital costs for the plants largely paid for, the U.S. nuclear fleet may be among the most competitive power generators available.

A New Look at Nuclear Power

Resolving this megatrend is critical to a sustainable energy future. Nuclear power could play a major role in solving the issues of the carbon/capacity conflict, reducing the rate of growth in U.S. CO_2 emissions over the next 20 to 50 years, and a tax on carbon emissions at a relatively modest rate could spur the investment necessary to increase nuclear power's contribution to controlling emissions.

A new generation of nuclear reactors offers the potential for safe, inexpensive electricity generation and hydrogen production. The pebble bed modular reactor incorporates a number of safety features that may make it more acceptable to a public and regulators. In the pebble bed design the nuclear fuel is encased in baseball-sized spheres that flow through a helium-cooled reactor vessel. Unlike currently employed light water reactors, the pebble bed design is immune to meltdown if the coolant gas is lost.

Another nuclear design, the high-temperature gas-cooled reactor (HTGR) promises generating efficiencies approaching 48%. This high efficiency would, in turn, improve the overall conversion potential for hydrogen produced via electrolysis. More importantly, HTGR is the only demonstrated nuclear technology that can supply the high-temperature direct process heat required for thermally driven hydrogen production processes. Results of a recent study indicate that an HTGR-driven steam-methane reforming process can compete with conventional hydrogen production at today's natural gas prices. Study results also indicate that hydrogen production via thermo chemical water splitting will be competitive with modest increases in natural gas prices and/or with economic incentives for reducing carbon dioxide.

Nuclear's Perception Problem

Obviously, nuclear power is controversial, and long-standing issues continue to fuel opposition to new nuclear construction. These issues include public perception of safety, the long-term storage of spent fuel and radioactive waste, and the risk of proliferation and diversion of nuclear fuel to weapons production.

Three Mile Island created a PR nightmare for the nuclear industry because of a lack of emergency communications planning and a media that failed to ensure knowledgeable sources or experts were given the opportunity to speak to the public and explain the reality of the situation. The accident was ultimately due to faulty sensing equipment, took 14 years and $1 billion to clean up, and did release radiation, but it never presented any danger to the surrounding areas.

This is obviously a hurdle for the industry and any possible investor. A more present issue, however, is the handling and storage of nuclear waste. The Yucca Mountain Nuclear Repository is now poised to accept spent nuclear fuel from 126 sites across the U.S. by 2017. There is much opposition to both the transportation and storage of this fuel underground. This will be an on-going issue for the industry until safety issues are accepted by the public.

More recently, the potential for terrorist attacks against nuclear facilities has heightened public concerns and prompted calls for increased security measures, including suggestions that any future nuclear plants be constructed underground. Resolving these issues is a political as much as a technical challenge.

Yet, smart money will follow nuclear, especially with the incentives from the 2005 Energy Policy Act. It is one of the main answers to get the trains of carbon conflict and capacity needs on parallel tracks. There is a movement of support from the environmental community—as painful as that may be to certain voices in the movement. Nuclear is the solution and opportunity in taking advantage of this megatrend, providing long-term balance and stability. That said, the political furor and potential public backlash will require patience and persistence.

Renewable Energy Sources

Renewable fuels, including wind, solar, biomass and geothermal energy hold promise for displacing fossil fuels. While historically expensive compared to conventional energy production, renewable generation is dropping in cost and growing at a faster rate than all other sources except natural gas. Record-setting venture capital and private investment is streaming into clean technology efforts. The Cleantech Venture Network tracked $2.9 billion in North American venture investment, an increase of 78% over 2005. This has put the clean technology industry on the map as the third largest investment category behind biotechnology and software. The federal government also anted up another $1.5 billion in 2006.

Interestingly enough, there appear to be more investment dollars than there are companies that can put them to good use. Yet, the investment has the ability to create economies of scale that can create viable markets for these technologies. Many companies that had been allocating investments to high-tech during the dot.com era, are now creating clean tech divisions such as GE's $2 billion wind unit and BP's solar program. And the prospects of a new clean tech job force has spurred economic development possibilities in many communities.

Renewables now only account for about 2.3 percent of the U.S. generating mix—up .3 percent from 2005. For renewables to make a more significant contribution, technological breakthroughs are needed to reduce costs, enable application in large scale and to more effectively integrate their operation with the existing power delivery infrastructure. Energy storage technologies, in particular, will help buffer the intermittent output of wind and solar facilities.

To date, 24 U.S. states and the District of Columbia have put standards in place that require electric utilities to generate a certain amount of electricity from non-hydro renewable sources. Most of these requirements take the form of "renewable portfolio standards," (RPS), which require a certain percentage of a utility's power generation to come from renewable sources by a given date (Figure 5-7).

Renewable Portfolio Standards

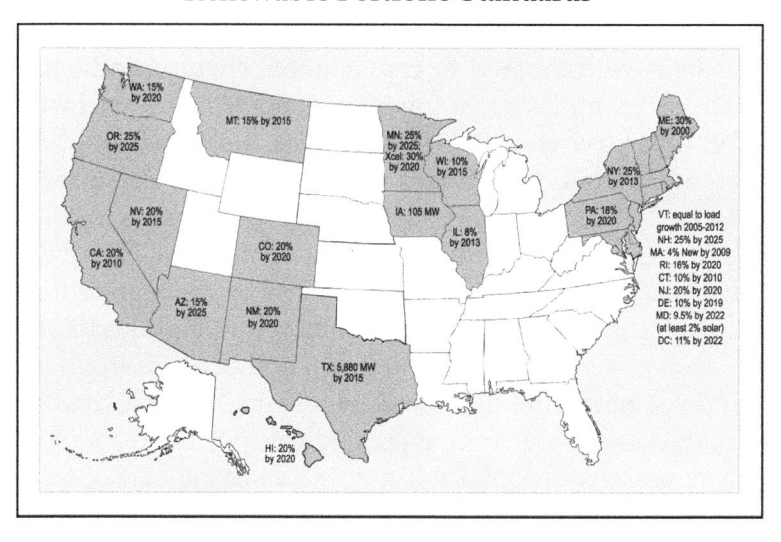

Figure 5-7. 25 states and the District of Columbia now employ voluntary or regulatory renewable portfolio standards for the implementation in a range of time frames. *Source: Pew Center web site on Global Climate Change*

State to state, there is a range of renewable power requirements and definitions of renewable energy vary. Standards range from 4% renewable generation by 2009 in Massachusetts to a more aggressive program of 25% by 2013 in New York. Three of the 23 states

(Hawaii, Minnesota and Illinois) have a voluntary program of renewable energy goals without enforcement provisions and many states allow utilities to comply with RPS through tradable renewable energy credits. While the success of state efforts to increase renewable energy production will depend in part on federal policies such as production tax credits, states have shown their considerable efficacy in encouraging clean energy generation.

While regulation can drive the implementation of cleaner generation sources, only when these technologies can sustain themselves in the market, are they successful and viable. There is clearly a place for renewables as the second favorite choice after energy efficiency as part of a broad and deep portfolio approach to energy supply and management. There are also key technical challenges—also representing opportunity—in balancing a system with a high percentage of renewable energy. In Hawaii, large injections of wind have had a negative effect on the system due to the variance. Hawaiian Electric Company has developed a "shock absorber." Experts in the field are not yet sure of how high a percentage of renewable energy will start to affect the operation of the electric power grid.

Regulation and technical challenges aside, investment in renewable power can help a company meet peak demand, gain a favorable green image, earn emission trading credits for emerging environmental trading markets and meet the challenges of the megatrend of the carbon capacity conflict.

It may be the combination of renewable resources and demand response that ultimately bridges the gap to assure supply from intermittent sources such as wind and solar. Of course, storage can provide an answer—but that promise has not been fulfilled as of yet even with decades of trying and millions invested.

Wind

Wind power has come of age and is gaining momentum. Over the past two decades wind power has developed into a cost-competitive generation option and the wind power industry is booming worldwide. Advances in wind power technology have reduced the cost of electricity from wind and improved turbine reliability.

Of all the renewable options, wind is the most practical large-scale generation option. Quite a few power companies are investing in this technology. Wind energy in the U.S. has soared from 1,600 megawatts in 1994 to more than 11,000 in 2006. The National Energy Renewable Laboratory expects that wind power will provide 20 percent of U.S. power generation by 2030. This positions wind power as a major component in dealing with the carbon/capacity megatrend.

The growth in wind generation brings new challenges. Wind generation is intermittent and can't be dispatched like most conventional generating sources, so large blocks of wind generation can complicate the scheduling of generation resources—and make wind output difficult to bid accurately even an hour ahead in forward markets. California, second leading wind generating state behind Texas, has about 2360 MW of wind generation, accounting for about 2 percent of total installed generating capacity. The California Independent System Operator (Cal ISO), which operates the state's electricity grid, must dynamically schedule transmission and other generation resources to respond to varying wind generation and balance total system load and generation.

Improvements in wind energy forecasting are expected to support system operation in cases where wind generation contributes more than a few percent of total generating capacity. The California Energy Commission and EPRI have developed a meteorological system for real-time forecasting of energy from wind turbines. Initial tests show the system is more accurate than persistence or climatology forecasts. Further research and development is necessary to achieve a more accurate and robust operational system.

While large energy storage capacity is an effective option for dealing with intermittency issues, the investment of energy storage capacity at the site of a wind farm of adequate size is typically too high. Alternately, a critical and core technology play in renewables is anchored around the concept of a system wide "shock absorber" that allows the intermittency of the source to be balanced in such a manner that the renewable is not allowed to create system wide distortions. These generally will consist of relatively small amounts of storage capacity for short-term, fast response situations. An installation at the Hawaiian Electric Company is designed to store energy during wind gusts and then return the energy to the grid during a lull or to compensate for sudden loss.

The other logistical difficulty for large scale wind production is location. Class seven, the top rated locations for wind—those with the strongest wind speed—create more, and therefore, lower cost energy. By doubling the wind speed from 12 to 24 miles per hour, a plant can produce eight times more power. Class seven sites are typically offshore where the wind blows the hardest. Areas in Montana Wyoming and the Dakotas have regions that are class six. Unfortunately, these more desirable locations are typically farther from existing transmission systems. Most wind development has been focused on land-based installations. There are many class-four and -five regions throughout the country that are closer to power transmission.

Offshore wind development is projected to be the longer-term solution to balancing wind power's benefits and in dealing with this megatrend. Two major commercial offshore wind projects have been proposed on the east coast for a total of almost 600 MW. Forward thinking companies will strongly consider opportunities for investment in wind installations.

Solar

Solar energy provides less than 1 percent of all energy generated on a global level and represents the smallest share of all renewable technologies for power generation in the U.S. However, solar manufacturers are experiencing exceptional growth and the popularity of this clean energy option is unprecedented. Although the price of solar is high, as fuel prices rise, climate change concerns increase and government incentives continue to encourage solar energy investment, the growing solar trend can only continue. The solar market, including all solar-related products and services, is now valued at more than $12 billion, but this number is expected to rise to $30-$40 billion by 2016.

Solar power has practical uses that are unique and make it a very flexible power source. Solar systems can be deployed in the most remote, grid-isolated areas, on hand-held calculators or other devices, on highway emergency phones, on residential housing and in large-scale power generation sites. On the other hand, there are limitations. The sun only shines 30 percent of the time, even in the sunniest locations, which makes it difficult to cover the cost of many installations with such a capital-intensive technology.

The photovoltaic industry is facing material shortages of solar-grade silicon. Due to growing demand and competition from the computer chip industry, high-grade silicon needed to make these products is in short supply causing prices to rise. Chip manufacturers are better able to pay higher prices, putting the solar industry second in line for product. With increased growth, the solar industry is forecast to overtake the chip sector in the next five years in its use of silicon. But it must first overcome these shortages of the commodity, which is hampering its expansion. This is a temporary issue, however, and as processors develop more capacity, shortages are expected to end some time between 2008 and 2010.

One solar technology that may provide a way around this materials shortage is thin film photovoltaics. Thin film solar panels use a very thin coating of silicon (as much as 99% less silicon than a solid silicon cell) or other photovoltaic materials such as copper compounds. The downside is that thin films are typically much less efficient than silicon, with 7% to 10% efficiency compared to silicon's average 15% efficiency, but efforts continue to improve this factor. Thin films normally take up twice the space of silicon panels with the same wattage but lend themselves to mass production and can be readily integrated into building products such as roof shingles, building facades and window glass. Many experts believe thin-film cells are the wave of the future, as they don't rely on conventional materials, are much less expensive and open many new markets for PV manufacturers.

Federal rebates for home and business installations are in place that reduce overall solar installation cost by 30 percent through 2007 and this may continue. Some states are also putting together their own incentive programs. In a 2006 California initiative, the public utility commission approved $3.2 billion in renewable rebates to be granted over 11 years. The overall goal is to generate 3 gigawatts of solar power by 2017, roughly 4 percent of California's total energy generation. On top of this, California put in place the Million Solar Roofs program which provides for an array of efforts to open up the market to solar including increasing the cap on net metering and making solar panels a standard option on all new homes.

These efforts promote the use of solar, although there is still the issue of cost. The cost of solar is still quite expensive—about $20,000 plus to equip a home with photovoltaic panels and $5-6,000 per kilowatt for a large-scale generation plant, five times that of natural gas plant and twice the cost of building a pulverized coal or IGCC power plant.

But with renewable portfolio standards in place and environmental issues on the forefront, large solar plants are multiplying and getting larger. At the end of 2004, the U.S. had only 397 megawatts of solar-energy capacity. It is estimated that 7,000 MW of centralized renewable power plants will be built by the year 2020, and possibly much more. In particular, solar thermal technologies, especially concentrating solar power (CSP) are gaining more attention. CSP technologies use reflective materials such as mirrors to concentrate the sun's energy and convert it to electricity and are much more cost-effective and practical then PV for centralized plants.

Current solar farms are generally in the 35- to 80-megawatt range, but things are looking up for these large-scale plants. Two separate CSP solar farms, are on the books for Southern California that, when complete, will be the largest solar projects in the world. The first, with the capacity to generate 500 megawatts of electricity will be built for SoCal Edison in the Mojave Desert, and the second, a 300-megawatt plant in the Imperial Valley, for SDG&E.

Solar can play an important role for companies dealing with this megatrend. Solar plants provide power during peak loads on hot days, just when it is needed, thus alleviating utilities from building other, less eco-friendly peak generating plants. This provides substantial opportunity for companies looking for key investments that can help to solve the carbon/capacity issue.

Biomass

Biomass is the most prevalent and widespread renewable resource world-wide and provides 10 percent of the world's energy needs. For thousands of years, people have burned wood to heat their homes and cook their food. China and Sweden get 19 percent of their primary energy supply from biomass fuels while India receives 42 percent. Sweden expects this number to grow to 40 percent by 2020. Some estimates forecast that biomass will provide 30-40 percent of the worldwide energy demand.

In contrast, in the U.S., biomass represents just 1 percent of the national total and one half of the non-hydro renewable capacity. The federal government and states are taking a serious look at this power technology and its share is expected to increase to 1.8 percent by 2030. Biomass has the ability to not only provide electricity generation for power providers and for industrial uses, but biofuels for use in transportation. Additionally, the fuel is different from other renewable sources as it can be generated and dispatched on demand.

Biomass encompasses a wide range of combustion technologies that use plants and plant and animal wastes as fuel. Feedstocks include grasses, combustible waste material, fast growing trees such as poplar and willow and methane gas captured from landfills or animal waste. Because most of this biomass residue will end up buried in landfills or burned in open fields, it is much more environmentally sound that these fuel sources be burned in a modern boiler equipped with emission controls. When biomass is produced and consumed on a sustainable basis, CO_2 released during combustion is equal to the amount captured by photosynthesis as it grew, therefore, there is no net carbon dioxide to the atmosphere. Additionally, biomass processes can reduce landfill.

There are many options for the use of biofuel. Cofiring biomass with coal is the primary use with existing technology. It reduces the amount of coal used while reducing CO_2 and other pollutant output and is expected to contribute almost 50 percent to the growth of biomass energy in the next 25 years. While the majority of biomass power is used for utility coal-fired power generation, it can be used for on-site industrial generation such as at pulp and paper mills. Gasification of biomass can also be used for gas turbine generation both simple- or combined-cycle with efficiencies approaching 40 percent.

Yet, there are issues. Gasified biomass must be used at the point of production and liquefied biofuels can be stored and transported, but the financial logistics of this can make it out of the question for most applications. Solid fuels are also expensive to transport which can limit the applications. While biomass power is one of the most affordable and accessible options available for rural areas and developing nations, cost is still an issue. A biomass plant can cost up to four times that of a conventional coal plant. But prices are dropping and estimates put the price at $1300 or less by 2010, competitive with the construction of a natural gas plant.

Biomass will be a key technology in navigating this megatrend. It is the most prevalent renewable fuel, and there is momentum in promoting its use, creating significant opportunity. The Energy Policy Act of 2005 aims to motivate businesses to increase biomass production by offering to provide funding to private enterprises for biomass and biofuel production. Many states have put programs in place to promote the use of biomass as well including Idaho, Colorado, and Montana.

Geothermal

Geothermal energy has been around for more than 35 years in California and as long as 70 years in Europe. This generation source accounts for about 16 percent of all U.S. non-hydro power generation, however, because geothermal plants are site-specific, companies face limitations in location. Also, the cost of some technologies are high and geothermal energy is expected to experience limited domestic growth.

Geothermal energy is much more prevalent in certain geologically active areas around the "Ring of Fire," particularly in eastern Asia and the western U.S. coast. Most world-wide growth will be in Asia, particularly in countries such as Indonesia and the Philippines where demand is high and government policies promote development

Regardless of its limitations, those areas appropriate for this technology offer opportunity for companies needing to comply with RPS or reduce their overall CO_2 footprint. In many markets, the cost of hydrothermal power is comparable to that of conventional electricity. And new technologies such as future flashed-steam plants are expected to experience significant reduction in the cost, primarily in well exploration and field development which traditionally account for one-half of a system's initial capital cost.

Other technologies include binary hydrothermal which requires an above-ground power plant making it almost twice the cost of the flash-steam method. Also used are hot dry rock systems which require extensive field development.

Hydropower

Hydropower provides about 6.5 percent of U.S. electricity generation and nearly half of the electricity produced in the Pacific Northwest. Hydropower is a renewable energy resource but is looked at separately as it is also considered a conventional generation source. Although hydro will continue to be an important source of low-cost renewable, non-emitting generation into the 21st century, its percent share of the generation mix is expected to decline to 5.3 percent by 2030. Most of the good sites for hydropower have already been developed, and any proposals for new dam construction would face public opposition as well as a burdensome licensing process.

Nevertheless, there are some opportunities for increasing hydropower capacity by repowering existing facilities without constructing new dams. Increasingly efficient turbines can generate additional power from the same amount of water. Additionally, of the 80,000 dams in the U.S., only 3 percent generate hydropower. The Department of Energy estimates that 5700 additional sites could contribute close to 30,000 additional megawatts to U.S. capacity.

Issues with hydroelectric plants include the harm to fish habitat and migration, yet new designs for turbines, screens and spillways and ladders can help fish migrate safely around these obstacles.

The opportunities with hydropower are limited for this megatrend, although those organizations with existing hydro assets will find that these will continue to play a key role in dealing with the carbon/capacity issue.

Hydrogen

Increasing concern with energy security and environmental issues is leading to heightened interest in hydrogen as a substitute for fossil fuels in some applications and as an energy carrier. It's important to understand that hydrogen is not a primary fuel source. Like electricity, hydrogen is a manufactured energy product, requiring significant energy input. In this context, the prospect for a "hydrogen economy" represents a significant expansion opportunity for electric utilities, either independently or perhaps in partnership with oil and gas companies.

Initially, hydrogen will be produced by the reforming of methane or natural gas, and from gasification of coal. In the future, hydrogen may be produced electrolytically on a large scale by renewable sources or nuclear power in a zero-emissions approach. This will only be achieved through considerable technological development supported by sustained investment in R&D.

Although touted as the fuel of the future, the challenge of hydrogen remains in its manufacture.

Distributed Energy Resources: A Paradigm Shift

While IGCC and nuclear units continue the legacy of the large central-station power plant model, distributed energy resources (DER) represent a fundamental change in the energy landscape. The DER concept encompasses small generators, energy storage devices and energy management systems located at customer sites or in the distribution system. DER technologies include fuel cells, micro turbines, and gensets for power generation; and flywheels and supercapacitors for storage. DER technologies range in size from 3-10 kW for residential systems to 50-500 kW for commercial users to 1-50 MW in the industrial market segment.

DER systems offer many opportunities in the evolving energy industry. DER has the potential to help energy companies capture new markets, serve high-value customers, reduce infrastructure investment, and optimize asset utilization. DER also holds great promise for improving the efficiency and reliability of the power system. Increased use of DER may delay or eliminate the need to build additional central generating plants and new transmission and distribution lines. Customers can use DER devices to provide standby power and to improve power quality and reliability.

Many people in the industry still use the term "distributed generation." This definition is both inaccurate and counterproductive because it limits our thinking about how to exploit DER's full potential. Using the term *distributed energy resources* is important because DER involves much more than generation. Properly planned and implemented, DER involves a systems approach to providing energy services. It's a mistake to think of DER simply in terms of small generators that provide backup power when grid power is interrupted.

The Opportunity:
Fuel Cells: Clean, Quiet, Compact Power

The real value lies in using DER to its full potential and applying it in combination with load control and energy management programs as a part of overall strategic planning activities to manage electricity demand. A leading DER technology, fuel cells, offers a good example. Fuel cells received a lot of attention and venture capital in the late 1990s, but after taking huge losses on the technology, the market soured. Stock prices plunged in 2001, and developers have had a difficult time finding financing to keep them going. However, now that the price of natural gas has gone up, the technology is turning heads once again. There are hurdles to making this a successful market and the technology is still in its nascent stage with few viable products. Yet the potential is large and fuel cells could play a key role in solving the conflict of this megatrend.

Fuel cells generate electricity cleanly and quietly by harnessing one of the simplest of chemical reactions: combining hydrogen and oxygen to form electricity, water, and heat (Figure 5-8). Fuel cells hold enormous potential in a wide range of applications that span the residential, commercial, industrial, and transportation sectors. Fuel cells can produce clean, quiet electric power for homes, commercial buildings and light industrial facilities—as a replacement for or, more likely, a backup to grid-connected power. In addition, fuel cells produce heat (cogeneration) which can be used for water and space heating or for industrial processes.

The Workings of a Fuel Cell

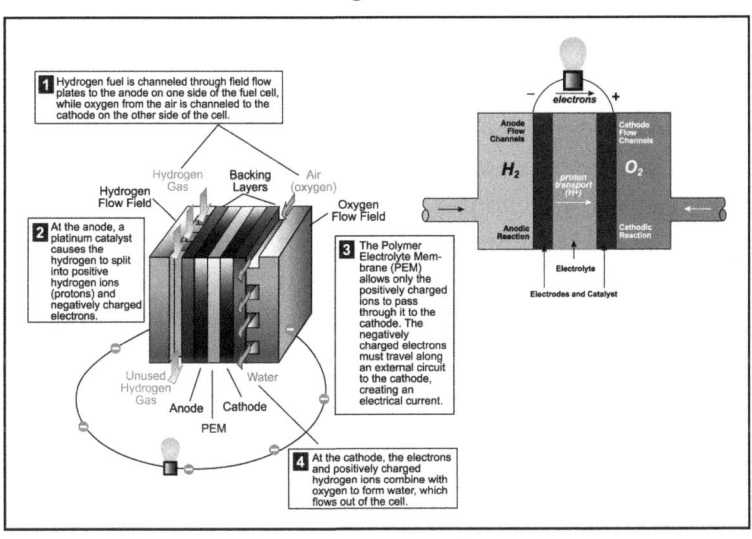

Figure 5-8. Through an electrochemical reaction, fuel cells combine hydrogen and oxygen to produce electricity, water, and heat providing highly efficient, compact, quiet and clean electricity for a wide variety of distributed applications. *Source: EPRI*

These attributes make fuel cells an attractive technology for providing power in crowded urban areas and in T&D-constrained locations. They could also provide combined heat and power (CHP, a.k.a. cogeneration) to buildings to increase overall energy efficiency, as well as peak-shaving support and load management.

This is especially attractive, for example, to commercial customers that absolutely need reliable high-quality power, yet also face pressure to control costs. Hospitals are a perfect example. Hospitals and similar facilities, and the energy companies that serve them, could exploit the potential of DER fuel cells with an *integrated systems approach* to ensuring power quality and reliability. A fuel cell serving the customer's facility and also connected to the grid not only ensures power quality and reliability, it can be part of a system that also makes money for both the customer and its energy service provider.

Most commercial building facilities managers don't have the time, budget, expertise or desire to purchase, install, operate and maintain a fuel cell system. It makes more sense for them to purchase those services from an energy company that's developing new business around the provision of high-quality emergency power. That energy company could use a combination of distributed fuel cell generation and energy management technology to pull off a real "killer app." The energy company and the customer would manage the on-site generation resource and the building energy usage in ways that allow excess power to be sold on the open market.

For example, if weather extremes drive up electricity demand and prices, the building's energy consumption could be reduced (say by dimming lights or changing the indoor temperature a few degrees) and the saved power can be sold on the market at a profit to the customer as well as the energy company. Similarly, if the customer did not use all of the fuel cell's output, the excess electricity could be sold. Or, if the customer used the fuel cell as the primary source of power, the unused grid power could be sold. In any case, the point is the same: the energy company invests in technology and deploys it as part of an integrated system that can yield multiple sources of revenue. In this example, the energy company sells not only the fuel cell, but also installation and O&M services, power quality and reliability, and energy management services as well as traditional grid power. Moreover, the energy company and customer form a business partnership that benefits both. A variation of this idea—using energy management systems and load control as a virtual power plant—is presented in Chapter 7, The Destiny of Customer Engagement.

The broad deployment of fuel cells and other DER technology will help provide a bridge to a future power system design that allows electron flow in all directions. (And, like IGCC, fuel cells will help pave the way to the future hydrogen-based economy. The DER fuel cells connected to customer power systems and to the grid will provide a market for hydrogen fuel and also provide the foundation for a hydrogen delivery infrastructure. A possible scenario involves the installation of DER fuel cells that would generate as well as consume hydrogen. During periods of low demand, hydrogen would be produced for use in refueling stations for hydrogen-powered vehicles. Thus, the use of hydrogen fuel in the transportation and electricity-using sectors would be complementary and mutually reinforcing, increasing the market pull and technology push for the hydrogen economy.

While much emphasis has been placed on the development of large-scale fuel cells, sufficient enough to power a house, car or supply a substation, the entry point to the market will more likely be in much smaller sizes to serve transportable loads such as computers, cell phones and related services. It may be a device for instant power and recharging that brings fuel cells into common usage. One could envision a fuel cell "home base" that ensures clean power for computing and entertainment needs (picture supporting a $10,000 home theater system) is used to charge laptops, PDAs and cell phones and not as primary, but secondary supply. This back up power might then serve residential load during peak times and become fully dispatchable by the retail energy provider in times of need.

And as discussed in Chapter 7, the distributed energy concept can extend to the level of individual consumers driving plug-in hybrid electric vehicles that can feed power back to the grid while parked.

Energy Storage: The Ultimate Answer

Energy storage perhaps represents the biggest single paradigm shifter in the electricity business and the "tipping point" opportunity for investment, deployment and new business opportunity. Whether it is through the use of ultra-capacitors, large scale promotion of hydrogen fuel cells or a return to pumped storage, the concept of building a supply balance point will bring the most dramatic change since the advent of nuclear power.

In the traditional power industry paradigm, electricity has been generated instantly to match consumption and delivered to the customer at the speed of light. Unlike other products, electricity can't be put in a box, placed in inventory, and then withdrawn when demand rises. As a result, power companies have little or no flexibility in managing production and supply. The ability to store electricity would be profoundly liberating and transformative. One obvious advantage is increased flexibility in meeting peak demand using power generated during off-peak periods. But storage promises to do much more in a strategic business environment.

It's important to view storage as an energy management technology, not as just another type of generation. In the wholesale power market, storage allows companies to buy electricity when it's cheapest rather than when customers most need it. The ability to stockpile cheap power strengthens a company's bargaining position in the competitive marketplace, and allows it to focus on buying or generating power at the lowest cost and selling to the highest bidder. Because storage plants are charged from the cheapest source of power, they help cushion companies from spikes in fuel prices or other uncertainties.

In the electricity supply system, energy storage provides a buffer that enables companies to more effectively manage the flow of power from intermittent forms of generation such as wind and solar facilities or from independent power producers. Storage also helps buffer customers from any lack of reliability or power quality from third-party generators.

Storage could also play a key role in alleviating transmission constraints. Installing a fast-responding storage plant near a transmission bottleneck or a load center may be an attractive alternative to adding new transmission or generating capacity. The storage plant could be charged during off-peak hours with inexpensive electricity and feed the power back into the system to serve peak load the next day.

The reason large bulk storage technologies have not been implemented in large scale in the U.S. is simple economics. Today, energy storage provides only 2.5 percent of total U.S. electricity capacity, most of it from pumped-hydro installations used for load shifting. In Europe and Japan, that number is 10 percent and 15 percent respectively due to better site characteristics and generation costs. The problem for the U.S. is that the scale and cost of these large systems requires a larger investment than building peaking combustion turbines for reserve generation capacity. But research continues in a number of demonstration sites to bring down costs, relieve grid constraints and offer greater system reliability.

Intermediate-scale storage technologies are becoming more attractive for transmission and distribution system operations. These technologies are focused on system optimization of existing T&D infrastructure and improving reliability for customers. They can provide ancillary services and enable utilities to defer more capital-intensive investment. This example also bridges the line to the destiny of customer engagement and offers profitable opportunities to companies that service the end-use customer.

It's also worth emphasizing that storage can operate on both sides of the meter. While larger scale bulk storage plants can help companies manage production and supply, smaller distributed systems can provide an array of value-added services. A battery or flywheel system installed near or on a customer's premises could protect against unexpected outages, price shocks, and deviations in power quality that could affect equipment operation.

Energy Storage Technologies

The electric power industry has evolved according to a fundamental principle: power must be produced instantly in response to customer demand—and delivered at the speed of light. Emerging energy storage technologies give power companies an energy management tool to stockpile electrons in inventory and withdraw them when they're needed. The ability to store electricity will have a transforming and liberating effect on the power industry, giving companies more flexibility in operating their systems, meeting peak demand, serving customers, and conducting power transactions in the wholesale market and supporting the expansion of renewable facilities.

The range of energy storage technologies includes:

- **Pumped Hydro.** Used for more than half a century, pumped hydroelectric storage represents the lion's share of installed energy storage capacity worldwide. Pumped hydro requires two storage reservoirs at different elevations. During off-peak hours, water is pumped from the lower reservoir to the upper reservoir. When electricity is needed, the water is released to flow through hydroturbines to generate power.

  Further U.S. development of hydro is limited. Most of the best sites are already taken by the existing 37 plants, and any proposed construction would face long lead times and environmental issues.

- **Compressed Air Energy Storage (CAES).** A CAES plant uses off-peak energy to drive a reversible motor/generator that compresses air into an underground reservoir (Figure 5-9). When needed, the compressed air is withdrawn, heated, and run through the motor/generator to produce power. The two CAES plants in operation today—a 110-MW plant in Alabama and a 290-MW unit in Germany—use underground caverns excavated from salt domes. CAES plants can also employ caverns mined from hard rock formations or in porous media such as aquifers or depleted natural gas fields—geologies found in about three-fourths of the United States, according to EPRI research.

Compressed Air Energy Storage (CAES)

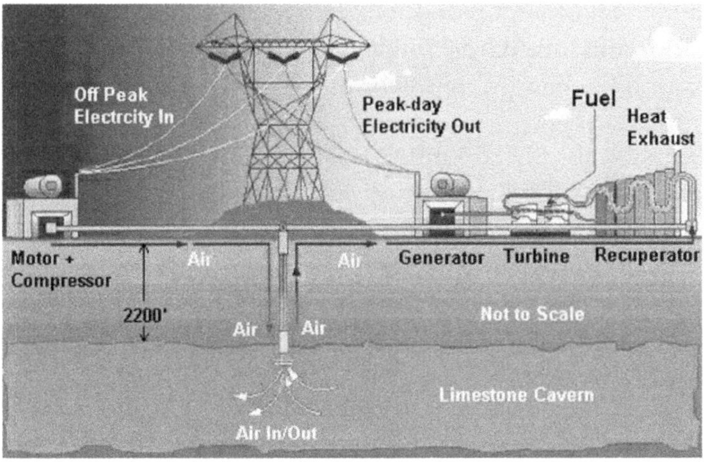

Figure 5-9. CAES is a fully proven technology with plants operating for over ten years, including a 290-MW plant in Huntorf, Germany, (1978) and a 110-MW plant in McIntosh, Alabama, (1991). These plants provide a variety of functions, including spinning reserve, load management, peaking power, and power factor control. *Source: EPRI*

- **Battery Energy Storage.** Battery storage offers a reliable and flexible means to improve power system performance and provide value-added customer services. Quiet and responsive, batteries can be sited near customers or in substations to provide backup power, grid support, or power quality assurance for sensitive loads.

   Advanced designs with high energy density and improved efficiency are outshining the familiar lead-acid battery. American Electric Power demonstrated a sodium-sulfur system that shaves peak load and protects power quality at an office park, while a nickel cadmium system is helping Golden Valley Electric Association stabilize the grid and reduce vulnerability to power outages in a remote Alaskan site.

- **Superconducting Magnetic Energy Storage (SMES).** SMES, which stores electricity directly in a donut-shaped electromagnetic coil of superconducting wire, is valued for the ability to store and discharge power at very high efficiencies and respond to changes in power output within milliseconds.

  Small-scale SMES units (less than 20 MVA) are already at work in utility distribution systems, where they provide power quality control and improve system reliability by injecting power strategically to provide voltage and frequency support. Large-scale SMES is currently constrained by the challenges of cryogenically cooling the system to superconducting temperatures.

  On the transmission system, integrating SMES with high-speed, electronic flexible AC transmission system (FACTS) controllers could prove a potent combination for improving the stability and throughput of constrained transmission corridors.

Incentives to Help Replace Conflict and Costly Compliance

A transformed, truly sustainable electricity sector would protect the environment while meeting society's needs for reliable and economic power. Energy policy and environmental policy would be coordinated, providing clear guidance for industry to meet agreed-upon performance targets and establishing realistic timetables for deploying new technologies and methods to reduce emissions and improve environmental quality. Power generation from all fuels would produce zero or minimal discharges to air, water and land, and the discharges that do occur would be offset by environmentally beneficial activities, such as ecosystem conservation or wetlands restoration projects.

The most notable feature of the sustainable future vision

will be the harnessing of market forces to resolve environmental issues. In sharp contrast to the all-stick, no-carrot approach taken by rigid regulatory regimes, market-based mechanisms and incentives will give companies both motivation and flexibility to voluntarily curb air emissions and more effectively manage waste streams and other environmental impacts.

Expanding Opportunities in Green Trading

There is no mandated federal CO_2 offset regulation today in the U.S. Fearing the discussion could drag on at the federal level, various state and regional voluntary, market-based programs are starting to emerge across the country in California, Oregon, and a nation-wide effort, the Chicago Climate Exchange. Meanwhile, seven northeast states have put together the Regional Greenhouse Gas Initiative with which they will implement the first mandatory cap-and-trade program for CO_2 emissions in the U.S. These programs have been a statement to the federal government and the public of a commitment by the electricity industry to CO_2 reduction. In fact, a group of industry and utility executives have recently gone before congress asking for a federal cap-and-trade program to reduce CO_2 emissions.

There are critics of these trading schemes from both sides of the aisle. Some will argue that mandated cuts are necessary to avoid a situation where pollution levels will concentrate in areas unable to cut their CO_2 output. Others argue that any mandatory cap-and-trade system would be costly and hurt the economy.

One key lesson from three decades of antipollution legislation is the importance of providing effective incentives to achieve the most efficient solution to environmental challenges. In many cases, command-and-control environmental policies can inhibit rather than encourage technology advancements. This results in reduced environmental progress, greater cost, and a focus on relatively inefficient "end of process" control strategies. An alternative to command-and-control is the use of market-based incentives to reduce environmental impacts including CO_2 reduction. Market-based incentives include emissions credits and offsets, as well as ecological asset management. In concert with the investments recommended in this and other chapters—including investments in technologies such as IGCC plants, the smart grid and energy/information portal—these environmental approaches can be an important part of a robust business strategy.

Emissions trading has already proven successful in the case of sulfur dioxide (SO_2), and the trading of CO_2 and other greenhouse gas emissions is likely to be part of any coordinated federal response to global climate change and this megatrend.

Allowing markets to work—and protecting their operation—may hold the key to balancing the environmental needs of society with its costs. This will allow for a host of new options, such as renewables and demand response, to be placed into play on a level basis with fossil based assets. Instead of artificially constructing financial incentives for efficiency, a proper functioning market can create an economic business model that hits the broad societal goal of cleaner air, reduced water use and more efficiency while minimizing cost. Rather than trading "bad" resources for "good" resources, a functioning market will allow a balance that is financially optimal and environmentally sound.

Ecological Asset Management

In a new approach to increasing revenues and reducing environmental compliance costs while protecting environmental resources, utilities can convert their landholdings into profitable economic assets. The key is identifying opportunities to participate in existing and emerging markets for carbon sequestration, wetlands mitigation, stream mitigation, and others. By converting ecological resources into profitable economic assets, companies can increase revenue, reduce environmental compliance costs, and remain active as environmental stewards protecting valuable natural resources.

This approach assigns monetary value to a category of natural resources called "ecosystem services." Ecosystem services are the raw materials supporting ecosystem and human health. They are the basis of economic productivity. Ecosystem services help regulate and maintain air and water quality, and help stabilize and sustain productive soil and biodiversity.

As the financial value of ecosystem services is established, they come to be seen as ecological assets, which have become tradable commodities. Ecological assets represent measurable improvements to ecosystem services, such as SO_2 and NO_x reduction credits, CO_2 mitigation credits, and waste load allocations, as well as less conventionally measurable attributes such as riparian or nutrient buffering, wetlands mitigation, stream restoration, aquifer recharge, and endangered species credits. These credits or allocations have become true commodities that are created, banked and traded—just like traditional financial instruments—in the environmental marketplace. This marketplace is poised to achieve tens of billions of dollars in transactions annually.

One example of market-based eco-asset management is a landmark agreement between Allegheny Energy, Inc., and the U.S. Fish & Wildlife Service that will preserve 12,000 acres in West Virginia as

Ecological Asset Management: A Systems Approach

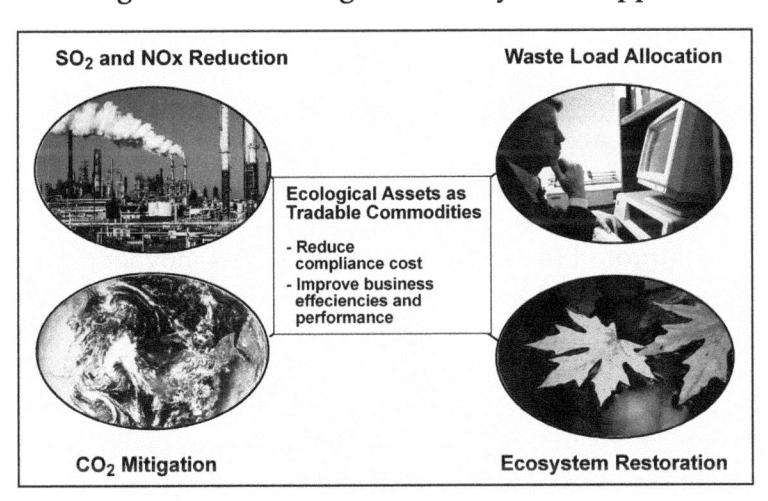

Figure 5-10. Ecological asset management takes a systems approach to assessing and managing environmental assets to increase land value, improve business performance and create environmental credits. The marketplace for ecological assets is poised to achieve tens of billions of dollars in transactions annually. *Source: EPRI*

part of the Canaan Valley National Wildlife Refuge. This complex transaction hinged upon a comprehensive appraisal of the property's fair market value, both in terms of traditional highest and best uses as well as eco-asset based values. An appraisal valued the entire tract at over $32 million, more than double the value from traditional uses alone. The eco-assets include mitigation credits associated with protecting wetlands, sensitive habitats and endangered species, as well as credits for sequestering carbon.

A systems approach to ecological asset management (Figure 5-10) could improve compliance cost and efficiency, enhance overall business performance and be a key part of energy company strategy in dealing with this megatrend.

The Case for Environmental Stewardship

Environmental policy has become a key driver of energy policy. Taking a proactive approach by investing in carbon reducing advanced technology helps energy companies not only reduce costs, but also shape the future of the industry rather than having it shaped for them by regulators and policymakers. Research in this arena yields new scientific and technical knowledge that helps companies inform regulatory and policy decision-making. This gives companies a "seat at the table" in policy deliberations to help influence the creation of regulations that are more flexible and include market-based mechanisms. Such regulations can reduce compliance costs, protect the environment, stimulate innovation and open new business opportunities such as environmental asset management, emissions trading, and green power services.

Better knowledge also translates into better business decisions and improved risk management. Environmental issues present complex challenges whose solutions only become clear when they're better understood. Sustained R&D on environmental issues opens a path from uncertainty to clarity, helping decision-makers understand the costs, benefits, and risks associated with different courses of action.

Public perception is also key. Fossil-fuel plants are high-profile sources of greenhouse gases and other emissions, while dams and transmission towers are clearly visible reminders of the environmental impacts of power production and delivery. Companies that take an innovative, forward-looking approach to environmental concerns can offset some of the negative perceptions. This commitment to environmental stewardship isn't just good public relations, it's good long-term business strategy. It orients a company toward new technologies, processes, and services. It appeals to environmentally conscious consumers who may be more receptive to new green energy programs. It reduces risk exposure, leading to higher stock prices and more favorable insurance rates. Moreover, it demonstrates to regulators and policymakers that companies are willing and able to take voluntary early action on environmental issues. This in turn helps support the development of less onerous, more flexible market-based environmental regulation.

Succeeding with the Destiny of Carbon Constraints and Capacity Demands

The challenges for regulated utilities are clear: how to gain recovery of costs for new generation technologies when their success is not assured. This is exacerbated by the threat that these new technologies may become the best available environmental control technologies (BACT), putting the industry in the position of needing to meet the same carbon and emissions footprints throughout the generation fleet. Even greater is the challenging probability that there will be a shortage of capacity—and utilities will be blamed for either condition—lack of power or power plants that are not clean enough.

This also holds true as well for public power which are looking at capacity shortages and will find the same challenges in securing funding for traditional generation resources. Public power will also find the requirements of EPACT extend to them. EPACT 2005 impacted public power far more than previous energy legislation. NERC reliability standards requirements, the mix of generation portfolios, the need to explore time of use rates and advanced metering are just a few of the areas public power must now follow, similar to its IOU brethren.

Overall, complexity in the business model due to the carbon/capacity conflict will increase, requiring a new kind of integrated resource plan—one in which demand response is seated equally at the table with supply-side resources.

How can today's power company realize the vision outlined above and navigate a path to a successful future? The answer doesn't lie in making incremental upgrades to existing equipment, but rather in taking light-year jumps toward technologies that will transform a business and confer competitive advantage. And, critically, if these technologies are to be successful there needs to be an integrative system that allows for the vagaries of the sources to be coupled with market mechanisms, technologically balancing generation in a seamless fashion.

In creating this new world, companies can work to ensure full monetization of generation solutions. Demand-side management options can only truly work when physically connected to the system and used as a counterbalance to supply-side resources. Reducing demand through efficiency alone does not allow for a true counterbalance of generation. Once a technological system is put in place to recognize DSM for its full value—and make it dispatachable—the balance can be achieved. A crucial factor to making this work is the freedom of existing and new market participants to play in this arena—taking the risk while being allowed the reward.

There is virtually no doubt that some form of carbon legislation will be put in place within the next three-five years regardless of who is in the White House or which party controls Congress. The national psyche has "voted" and continues to pressure for these changes. The Supreme Court ruling in April, 2007 placed accountability for carbon regulation with the EPA—even if the EPA was an unenthusiastic recipient of that honor. The 5-4 decision in the court's first-ever case on global warming forces the EPA to re-evaluate whether its regulation of tailpipe emissions should include carbon dioxide. The ruling could also lend important authority to efforts by the states either to force the federal government to reduce greenhouse gas emissions or to be allowed to do it themselves such as California's AB32 which is designed to cut carbon dioxide emissions to 1990 levels by 2020 and New York is leading an effort to strengthen regulations on power-plant emissions.

Making the right assumptions will be critical in developing a successful generation strategy.

- **Take the long view**—commodity prices and demand forecasts fluctuate much more quickly than plant additions and deletions

- **Maintain a balanced portfolio**—fuel mix should change with time, but not widely to protect against price and regulatory fluctuation

- **Be diligent in fuel assessment**—stay aware of energy and fuel market volatility and maintain vigilance against fuel blindness

- **Anticipate the carbon conflict**—stay on top of possible environmental regulation that can direct generation choices and create opportunity

- **Invest in technology**—use new technology to your advantage to improve operational productivity and in anticipation of new environmental regulation

- **Consider wild cards**—the economy, advanced technology, energy efficiency

Where to go from Here

This megatrend offers significant advantage to incumbent players who either have large coal generation capacity, or are willing to make the investment in clean technologies or CO_2 reducing systems. As with all the megatrends, technology and vision are at the core of a company's ability to capitalize on these opportunities.

While we have provided many technology solutions that can provide opportunities to work with this megatrend, strategic planning which incorporates several of these investments within the structure of competitive regulation, can lead to significant business growth and protect the market space available to these utilities. Some suggestions are:

- A balanced portfolio of core generation capacity in nuclear, coal and natural gas will be required moving into the future. Coupled with active demand response and load control, this type of new integrated resource planning, focused just on lowest cost will become the norm.

- Risk mitigation will require investing in a suite of fuel sources. Hesitancy in acting will not serve customers nor, in the case of investor-owned utilities, shareholders. Relying on the market to provide generation is a risky approach. As the industry saw in the early 2000s, leaving the market solely to the natural gas spark spread results in financial dislocation, regulatory wrath and an inability to pass prices to customers quickly enough as to avoid financial problems.

- The use of renewables such as large scale wind generation, backed by coal will require significant investment in the transmission system—as well as major political battles. California is imposing a restriction which theoretically would make it impossible to use wind backed by coal—and this will certainly be tested through the balance of the decade as demand grows and supply gets constrained A major opportunity for renewables lies in the connection with demand response as virtual "batteries," allowing the fluctuations in supply to be balanced with DR resources. There is a significant opportunity in investing in systems and technologies that allow for the intermittency of renewables

- Clean coal and carbon capture technologies will require a level of risk that some in the industry are yet unwilling to take. However, there will be those foresightful companies who make the investment and will earn long term rewards. It may require a new set of entrepreneurs who see the potential conflict as an opportunity for investment as the market for carbon is established.

Who should pay for all of this? Taking these ideas into the Competitive Regulation Investment model allows for a balance of low risk, fixed return, shared risk with a medium return and high risk with the possibility of extremely high returns. This balance provides the power provider with to deal with the destiny of the carbon capacity conflict. If the industry is to succeed it will certainly need regulatory support. The suggested investment model would allow for rate-basing fundamental pieces of the industry to ameliorate risk (such as Federal accountability for ownership of carbon once it is injected in the ground or underwriting the cost of carbon trading) while stimulating the marketplace to work.

Similarly, investments in LNG terminals, support of new nuclear and transmission construction and siting all have components of a new regulatory and market compact. This is true whether it is a collection of municipal utilities or the largest investor owned companies. The destiny of carbon/capacity conflict has an impact across the market and will change the dynamics of the industry for the next 20-30 years.

Investments that enhance operational effectiveness are funded at the next level. Risk is shared between regulators and shareholders and returns are based on environmental performance. Cleaner operations mean fewer emissions credits bought or sold. Ecological asset management efforts may be funded at this level.

New environmental service opportunities funded at the top level allow for expanded revenue. Green trading of emissions credits fall into this category, as well as strategies for enhanced environmental dispatch.

Environmental Investment Model

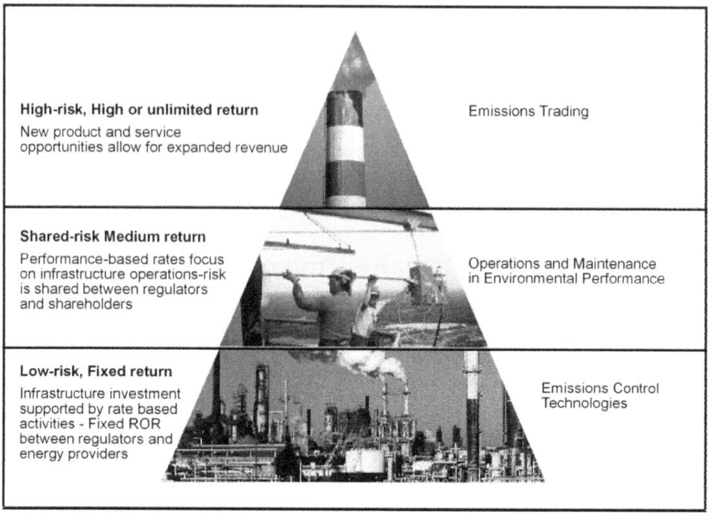

High-risk, High or unlimited return
New product and service opportunities allow for expanded revenue — Emissions Trading

Shared-risk Medium return
Performance-based rates focus on infrastructure operations-risk is shared between regulators and shareholders — Operations and Maintenance in Environmental Performance

Low-risk, Fixed return
Infrastructure investment supported by rate based activities - Fixed ROR between regulators and energy providers — Emissions Control Technologies

Figure 5-11. Rate-based investment in the environmental arena includes enhanced emissions control technologies. Moving up the triangle, performance–based activities may include environmental performance improvements through investment in operations and maintenance. Higher risk investment would involve green trading of emissions credits and other activities that provide for expanded revenue.

In this model, as applied to power generation (Figure 5-12), basic infrastructure investment is supported by rate-based activities under the traditional fixed rate of return arrangement between regulators and energy providers. Base load nuclear and fossil plants have traditionally been funded through this level, which offers a low-risk fixed return. Advanced fossil and nuclear technology such as IGCC and pebble bed, and renewable generation, can also be supported under the rate base or under performance-based rates where risk is shared between regulators and shareholders. Primary activities that would be covered under performance based investment would revolve around operations and maintenance such as the reduction of emissions, increased VAR output or flexible dispatch. The more efficiently plants are run, the better the return.

Power Supply Investment Model

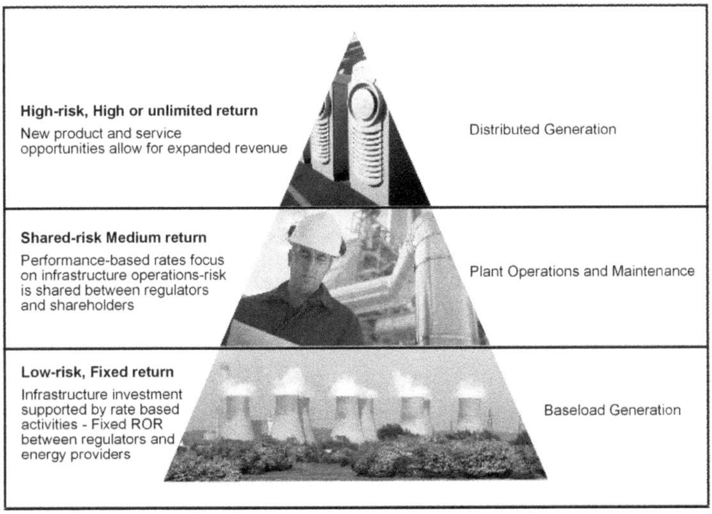

High-risk, High or unlimited return
New product and service opportunities allow for expanded revenue — Distributed Generation

Shared-risk Medium return
Performance-based rates focus on infrastructure operations-risk is shared between regulators and shareholders — Plant Operations and Maintenance

Low-risk, Fixed return
Infrastructure investment supported by rate based activities - Fixed ROR between regulators and energy providers — Baseload Generation

Figure 5-12. Rate-based generation investment provides a financial and operational foundation in base generation plants and other infrastructure. This foundation provides for further investment that balance risk and return in performance-based activities in Operations and Maintenance and higher-risk investments in technologies such as Distributed Resources.

Moving further up the model, new service opportunities allow for expanded revenue. Advanced technologies like fuel cells deployed in combination with energy management systems, or other high-performance distributed generation or supply technologies, such as energy storage, that could provide clean and efficient high-value power could be funded at this level.

A system that is at once sustainable, environmentally sound, and profitable is not a pipe dream, but a practically achievable goal. The pathway to that power supply destination is marked by sound investments in advanced technology that balance risk and return.

References

Summarized from *Insull*, Forrest McDonald, University of Chicago Press,

1962
Annual Energy Outlook 2006 and 2007, Energy Information Administration
"Nuclear Power's Role in Meeting Environmental Requirements," EPRI Technical Report 1007618, January 2003
"Industry Statistics," Edison Electric Institute
"Electricity Sector Framework for the Future," EPRI, 2003
"Putting Wind on the Grid," *EPRI Journal*, Spring 2006
"Nuclear Power's Role in Meeting Environmental Requirements" EPRI technical report *1007618*, December 2002.
"The Coal Option in a Boom-Bust Market." Mark Gabriel, presentation at the North American Energy Strategies Conference, April 7, 2003.
Wall Street Journal Energy Diary: King Coal
Sean Captain, "Turning Black Coal Green," *Popular Science* 2/2/07
Ellison, Katherine and Daily, Gretchen. "Making Conservation Profitable," *Conservation in Practice*, Spring 03, vol. 4, no 2, pp. 13-19
"Renewable Energy for a Sustainable World," EPRI, 2002
"Renewable Portfolio Standards Map," Pew Center on Global Climate Change
Energy Biz Insider:
- "Biomass' Prospects," *10/21/05*
- "Solar Rides Wave," 5/10/06
- "Solar Getting Limelight," 3/07
- "A Gust of Wind," 4/28/06
- "Clean Tech Goes Mainstream," 6/06
- "Offsetting Carbon Emissions," 4/7/06
- "FutureGen Advances"
- "Three Mile Island in Retrospect," 2/16/07
- "Energy Storage on the Horizon," 1/06

Utilipoint Issue Alerts:
- "New Nuclear Plants Coming to the U.S.?" Bob Bellemare
- "Fuel Cells Receiving Renewed Attention Due to Rising Energy Prices," Christopher Perdue

Chapter 6

The Destiny of Intelligent Infrastructure
A Megatrend of Connectivity

America's electricity infrastructure is ill-equipped to sustain our country's needs today, and wholly insufficient to handle the growth in demand that is projected over the next few decades.
Former US Energy Secretary Spencer Abraham
May 8, 2002, upon releasing the DOE's National Transmission Grid Study

Samuel Insull pioneered new business concepts and technology development including AC/DC conversion for improving transmission capacity and the expansion of the power grid for rural electrification. Insull not only accepted change, he demanded it. "In my business," he said, "the best asset is a first-class junk pile," and his junk pile was heaped with obsolete ideas as well as obsolete machines.
Forrest McDonald, *Insull*

The concept of the intelligent grid is not new, yet it is perceived by most as a "vision of the future" rather than a megatrend in our midst. This perception is both true and false on different accounts.

Our perception is false in that development of this intelligent system is happening now, incrementally, in small and big steps by a majority of companies. Implementation of distribution and substation automation, outage management and automated meter reading, are all leading to a digitally enabled, self-healing power network that provides affordable electricity to all classes of customers through ensuring the security, quality, reliability and availability of power. It will take time to evolve into a truly and fully integrated intelligent grid, yet it is happening now as a megatrend of large potential impact—$28 billion will be spent in the next five years on transmission and distribution projects, much of it adding intelligence to existing systems

Our perception is true in that companies are challenged in their ability to adapt to and take full advantage of digitally intelligent system operations. This ability requires companies to think about management and operations in a new way. The constant flow of information coming from intelligent networks means a company can operate in real-time, using real data—enterprise wide—and this will require process reorientation. Companies need to understand their processes and be willing to reinvent their company's order to gain the full impact of operational efficiency and all the many benefits that are delivered by a fully integrated intelligent system. The sooner the better—those companies that can embrace these changes and adapt to the destiny of intelligent infrastructure will be the companies that excel in the coming decades.

One of the interesting challenges is in definition and accountability. A recent Google search yielded 400,000 different web references to "smart grid"—with dozens of definitions as it is being co-opted by each constituency from the green movement to the Department of Energy. And, by implication, smart grid is described as a singular "thing" as if it will replace the existing grid in one swift motion. There is also an implied tone from some that distributing the power of the grid is the answer to a better network—but as we have seen from the transportation sector—clearly distributed resources (i.e. cars)—that is not always the most environmentally benign or cost effective.

The trend towards development of a more sophisticated and reliable network is occurring across all energy markets enabled by communications, the microchip and advanced computational tools. From smart oil fields which analyze flow rates, spots leaks and automatically balance CO_2 injection to the IntelliGrid that automatically phase shifts, provides adaptive islanding capabilities and projects cable failures just prior to their occurrence, meshing intelligence with physical systems will put significant new demands on companies in exchange for powerful levels of control.

The reality is a more practical vision of an intelligent infrastructure—that of the emerging electronically integrated energy provider that bridges information technology with system operations, power plant dispatch and trading all the way to the customer end of the business. For these systems to work, an extreme amount of native intelligence is needed in all devices, across all platforms and through each operating system.

Managing the enormous amounts of system data will require levels of artificial intelligence as companies move from monthly reads of meter data to 15-minute intervals—12 vs. more than 35,000 per year. Analyzing loop flows in transmission and automatically determining injection points for VAR support, coupled with intermittent renewable resources mean the systems will be able to manage a more diverse generation and demand side portfolio. These changes will also result in a new way of thinking and managing that will require process reengineering, significant job shifts and the need for more powerful computing. It will also require changes in thinking as systems get smarter further out on the grid, changing the need to drag information back to the core for decision making.

While regulated businesses have been waiting for regulatory blessings on systematic additions (at least publicly), most have been quietly making changes to their systems in a phased approach. And the provisions in the Energy Policy Act of 2005 have helped to spur both transmission and distribution development and provide more certainty in approval processes and rates of return. EPACT put an emphasis on system security and integrity, and set the path for state regulated utilities to seek returns for those investments. Similarly, EPACT has also stimulated municipal and cooperative systems to consider and act upon its recommendations for more advanced infrastructure. The rapid advent of automated metering infrastructure—or at least the pressure to make it occur—will forever change the way the system operates. These tools and technologies are necessary for efficient operations and will become the de facto standards if companies are to play in a broader context.

EPACT changes and additions now making their way through Congress contain numerous provisions to stimulate intelligent infrastructure investments. One recent version had $500 million set aside in support of smart grid applications as well accelerated depreciation on meters down to five years.

The expansion of regional transmission organizations will require bringing more intelligence to the electrical network. Companies involved in actively trading and manipulating their natural gas supplies will be forced to bring their systems up to standards that have yet to be established. At the same time, whole classes of smaller energy providers run the risk of being left behind as surrounding areas move further into the 21st century—and left with significant costs as they play catch-up across a customer base that is too small to support these upgrades.

Yet, the benefits of intelligent grid investment are clear. In a recent report published by the Energy Policy Initiatives Center, a scenario of smart grid implementation on the SDG&E electric grid is outlined. The study shows an initial $490M investment in thirteen improvement initiatives would generate $1.4 billion in utility system benefits and nearly $1.4 billion in societal benefits over 20 years. Utilities working with this megatrend through smart investments in intelligent transmission and distribution technology will reap benefits for themselves, their shareholders and their customers.

An interesting aspect to this megatrend is its interplay with the destiny of demographics. One way to ameliorate some of the aging workforce challenge is in the development of new classes of skilled workers that can produce and operate sophisticated tools and technology for system operations. From production and delivery to system management and consumer interfaces, industries undergoing revolutionary changes such as this will require and are dependent upon a skilled and knowledgeable workforce.

Our National Power System... Now

The national power grid is often described as the largest and most complex machine ever built. It is a vast interconnected network of high-voltage lines, control centers, sensors, substations, and distribution lines—all operating in concert to balance the supply and delivery of power instantaneously in response to consumer demand. Coupled to this physical infrastructure is a data network that supports the safe and reliable operation of the power grid and the management of power transactions.

The grid evolved over the past 100 years, through the piecemeal growth of regional islands of electrification served by individual utilities. Gradually the utility service territories were interconnected to enhance reliability and enable neighboring utilities to buy and sell small blocks of power. Aggressive technology development and bold infrastructure investments made it possible to bring power from remote generating stations to far-flung businesses, factories, farms and homes. As a result, electricity became the great enabler of technological progress, social transformation, and economic growth.

Today the power grid is facing challenges it was never designed to meet. This chapter will describe the challenges facing the power delivery system, the technological advances that can help businesses meet the demands of this megatrend, and a strategy for making sound investments in grid technology.

Reinventing the Grid:
20th Century Technology, 21st Century Demands

Despite incremental upgrades, the nation's transmission and distribution system has not evolved to keep pace with technological trends and the growing demand for electricity. Today's aging and congested power grid is becoming less an enabler of innovation and prosperity as much as it is becoming a drag on productivity growth.

U.S. demand growth in the next ten years is expected to increase by 19% while transmission expansion over the same period is projected at 6%. As a result of this imbalance, many parts of the transmission system are operating close to their thermal and stability limits. The move toward deregulation of wholesale power markets has increased the challenges for transmission operations by increasing the number of independent power producers and long-distance bulk power transactions. Electrons follow the path of least resistance, not necessarily a straight transmission path between supplier A and buyer B. The current flows over various routes between source and destination, resulting in parallel and loop-flow problems as grid traffic increases. This has led to grid congestion and bottlenecks that impede the efficient movement of power, reducing its reliability and its ability to support wholesale power transactions. Add to that the growing addition of renewable energy sources and other distributed energy resources, and the complications of planning for T&D growth only get more complex.

Designed to supply power to incandescent lights, simple motors and analog devices, the current power system still relies largely on 1950s-vintage electromechanical switching to supply power to end users. And increasingly, those end-use loads include digital electronics, computer networks and microprocessor-controlled equipment, some operating critical loads for hospitals and community emergency response. The risks are high yet the solutions are available for utilities and entrepreneurs that follow this megatrend of connectivity rather than the many market trends that abound.

SQRA: Power Provider Measurement of Choice

The business drivers and overall benefits achieved through the intelligent grid are dependent on SQRA—security, quality, reliability and availability—from the availability of high-quality power to secure protection against terrorist attack. In reality, the greatest impact on the cost of energy is not the price of electricity and gas, but the cost of power SQRA.

Power Security

Security is one of the most important issues that utilities face. There are really two discussions on security as they relate to the intelligent grid: external security of critical infrastructure systems and operational security of the electric system controlled by the intelligent grid.

The sheer size and complexity of the grid raises security issues. The grid is vulnerable to physical damage from natural disasters such as storms or earthquakes which fortunately the industry is highly experienced in dealing with. Man-made threats, however, including vandalism, terrorism and accidents present more of a challenge to understand how to deal with. Yet the most urgent security threat for utilities to understand may be that of cyber attacks to on the computer hardware and software systems that manage grid operations and process data in support of financial and power transactions. As with the 2003 Northeast/Midwest Blackout, the lack of real-time information left operators wondering if the cause was from a terrorist attack, accidental outage or system failure

Power Quality

Currently, about 12% of the electricity generated flows through a microprocessor-controlled device. That share could jump to 50% by 2020. The steady proliferation of digital technology creates challenges for the power grid. Digital devices are sensitive to power quality disturbances such as voltage sags and frequency deviations. For many consumers operating electronic equipment—chip makers for example—even a split-second interruption in electric service can crash sensitive equipment and turn a production run into an expensive mess.

In addition to raising power quality issues, the digital revolution is also beginning to change the power system's behavior. Microprocessor-controlled devices not only are sensitive to minor voltage sags that are normal in utility-grade power, they can also reintroduce electrical disturbances back into the power grid. This makes the grid's operation less predicable, so it must be operated with more conservative safety margins at the expense of peak performance and efficiency.

Power quality issues arise from the interaction between the transmission and distribution systems as each system impacts the other. The distribution component has been often overlooked in the total solution of power quality problems. Distribution systems have the means to improve the stability of the operating point or take a heavy economic toll. A 2001 EPRI/Primen study shows the total annual cost of power outages and power quality disturbances on U.S. businesses at an estimated $120 billion—against total retail electricity sales of $220 billion (Figure 6-1).

Power Reliability

The most critical business issue facing industry CEOs today, worldwide, is improving reliability according to the Black & Veatch *2006 Strategic Direction in the Electric Utility Industry*. Clearly, the key benefit is improved customer service, but the positive financial impact reliability can have for the company and for society can be just as critical. The cost to the U.S. of poor power reliability equates to an August 14, 2003 Northeast blackout every 2 weeks.

Even without outside interference, electric power reliability

Annual Cost of Power Outages and Disturbances to the U.S. Economy

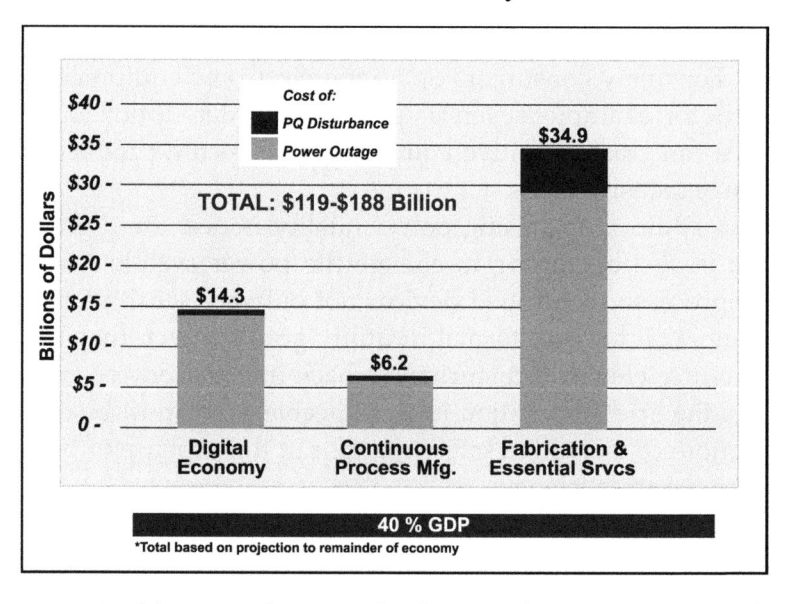

	Digital Economy	Continuous Process Mfg.	Fabrication & Essential Srvcs
Billions of Dollars	$14.3	$6.2	$34.9

Cost of: PQ Disturbance / Power Outage
TOTAL: $119-$188 Billion
40 % GDP
*Total based on projection to remainder of economy

Figure 6-1. In this 2001 Primen study, the cost of Power outages and disturbances on the U.S. economy is estimated at $119 to $188 billion per year—40% of the current GDP. *Source: EPRI/Primen Study: The Cost of Power Disturbances to Industrial & Digital Economy Companies*

problems have led to more blackouts in recent years than historically experienced in North America. It is clear that the reliability of the nation's electric power infrastructure has deteriorated and a continuation of the decline in the quality and reliability of electric power will seriously impact the nation's economy, as it affected California in 2000 and 2001.

Demonstrating the impact of poor power reliability in a single event, a Mirifex/REI survey regarding the August 14, 2003 power outage in the Northeast and Midwest, 66% of the businesses surveyed incurred a full business day of downtime while 25% were impacted for two or more business days. Over 20% lost more than $50,000 per hour of downtime, or at least $400,000 for an 8-hour day, while one business in ten lost between $100,000 to $500,000 per hour. 3.5% of the businesses surveyed lost more than $1 million for each hour of downtime. This is a pretty significant cost to the U.S. economy based on one isolated event.

The under-investment in needed infrastructure to meet growing customer demands is the culprit. The destiny of intelligent infrastructure is a direct reaction to this and the other critical power delivery issues the industry is facing.

Power Availability

The quantity of power supply and control of that supply is becoming an increasing challenge from both potential limitations of supply due to capacity challenges and the issue of congestion facing many critical areas. The assumption that sufficient supply means sufficient availability is no longer the case as system constraints continue to mount. An intelligent system is needed to ensure availability, providing such features as adaptive islanding, fault anticipation, re-routing through congestion and other trouble spots.

There is a close linkage between reliability and availability in that each is a measure of the time that power is flowing to an end-use process and that process is up and running. However, availability is a much more relevant metric to the customer as it acknowledges that once a process in interrupted, time is usually required to make repairs and get things up and running again. There is distinct difference between the availability of power in a single one-hour power interruption than that of sixty one-minute interruptions. This is an important distinction for customers with sensitive digital loads such as data centers or hospitals and other critical services. The time necessary to restart operations can significantly impact productivity.

Breaking Investment Gridlock

In the face of these challenges, the current grid is increasingly unresponsive, vulnerable and obsolescent—a 20th century infrastructure that's inadequate to meeting 21st century needs. And yet this infrastructure represents a $360 billion asset—a national treasure—that cannot be abandoned or entirely replaced. But the grid is overdue for a technological makeover that will enable it to handle large power transactions, power the digital economy and enable the next wave of technological advances.

Despite the dire need for grid improvements, the current regulatory, political, and business climate present obstacles to investment, especially in the transmission sector. The organization of the restructured transmission industry has created confusion about the ownership and operation of transmission assets, and the resulting uncertainty serve as a strong disincentive to investment. At present, the ownership, operation and management of the transmission system is performed by different organizations. Ownership of the lines remains largely in the hands of the traditional vertically integrated utilities, although some transmission companies (Transco's) have formed, mostly as passive owners of assets controlled by system operators. Operation of the system is in the process of being turned over to independent system operators (ISOs) and regional transmission organizations (RTOs). Meanwhile, there are also utilities and transco's that both own and operate. To further complicate the situation, the transmission system is regulated at both the federal and state level, and the direction of policy and jurisdictional control remains in flux.

As a result, owners of transmission assets are reluctant to invest in their assets when regulators may require them to be operated by independent transmission operators with points of view that may conflict with the owners' interests. Until the rules of the game are fully clarified and transmission owners gain some assurance from regulators that they will recover their costs, they will continue to squeeze their assets to stay in business and serve their customers.

One result of the dramatic deficiency in annual transmission investment has been bottlenecks in the power delivery system that impair economic growth and productivity and drive up the cost of electricity in selected areas. According to the Federal Energy Regulatory Commission, transmission bottlenecks cost consumers more than $1 billion in the summers of 2000 and 2001 alone.

The Edison Electric Institute estimates that investments in transmission assets in 2005 were $5.8 billion and close to $40 billion in the last decade. Compare that with roughly $200 billion in generation investment. This 5:1 ratio seems to be reversing itself, yet while a considerable amount of investment has already been deployed in new transmission initiatives, much more will be required.

Traditionally, only utilities would invest in transmission. However, through the last number of years of deregulation and restructuring, it is clear that while utilities are essential to the transmission business, independent developers are essential as well due to their innovation and willingness to take risks. Investment from both utilities and merchants will be necessary to meet the coming delivery needs of electricity demand.

Investment in distribution upgrades has been equally as difficult. While utilities spent more than $14 billion on distribution systems in 2005, they struggle to make necessary intelligent system upgrades with tightened budgets, other critical priorities and the budgets needed for distribution growth. To meet replacement, improvements and the requirements of system growth, the $14 billion spent in 2005 will need to be increased substantially in the coming years.

All utilities fight to bring balance to the need for investment in infrastructure with the desire to keep costs to customers in line.

After years of capped rates and, therefore, capped expenditures for IOUs, the systems are in desperate need of upgrades, yet this comes at the same time of rising energy prices, demands for carbon and emissions control and the demand for high quality power on the rise. Political and regulatory pressures, while trying to protect the customer, have kept utilities from making the investments that would normally be considered prudent. There is almost a perverse logic in play: utilities hesitate to make investments in infrastructure which undermines reliability for fear of non-recovery, upsetting customers and regulators. When improvements are made or proposed which result in rate increases, customers, especially industrial concerns, unclear on the benefit to service will typically fight any increases. Public power faces the same challenge in escalating costs while the improvements in distribution and transmission networks could allow for greater efficiency, lower loses and better power quality.

Breaking investment gridlock requires political rather than technical solutions. Nevertheless, beyond the investment barrier awaits an array of technology-based opportunities for reinventing the power grid to support 21st century society.

The Vision: Enhanced Power Flow with Digital Control

"The best minds in electricity R&D have a plan: Every node in the power network of the future will be awake, responsive, adaptive, price-smart, eco-sensitive, real-time, flexible, humming—and interconnected with everything else."

Steve Silberman, *Wired Magazine*, July 2001

The digital microprocessor-based technologies that are creating challenges for the current power grid are also the key to its revitalization. The present system's 1950s-era electromechanical controllers are too slow to govern the flow of alternating current in real-time, aggravating the problems of loop flows and bottlenecks.

Replacing the old devices with solid state power electronic-based controllers can increase the amount of power that can flow on transmission networks and enhance overall system reliability. Early application of solid-state controllers is paving the way toward a unified smart grid of the future that will move large blocks of power precisely and reliably while managing a growing number of commercial transactions.

The intelligent grid holds potential not only for solving the current system's problems, but to provide a new power delivery infrastructure to support the digital economy and a new generation of energy products, services, and end-use devices. Most of the technologies to enable this vision are emerging or already available. It is the integration of these technologies into a coordinated system that needs to take place. As incremental steps are taken within this megatrend, this coordinated system will emerge—the intelligent grid.

Advanced Metering Infrastructure

There is no intelligent grid without intelligent metering systems and data management. Advanced metering infrastructure (AMI) and meter data management systems partner to represent the linchpin of the system as they provide streams of actionable data that can be used to actively manage and monitor systems and customers. While the spread of automated meter reading has accelerated over the past 10 years, it is the movement to systems which include computing intelligence and communications that are making the difference. Companies which installed drive-by meters 10 years ago are looking at a technology refresh in order to maximize the potential for the systems. These technologies are a bridge to a future where the successful power company will one day engage its customers through a two-way energy/information portal which will enable a fully functioning marketplace with consumers responding to price signals in real time and engaging in a host of services.

Advanced meters now can provide the window into which system operators can understand the dynamic nature of what is actually occurring. Gone are the days of system estimation by past experience with real-time information being deployed on a massive scale. Whether it is outage management, phase shifting, load forecasting or fault anticipation, AMI brings the system to life. The destiny of customer engagement comes into play here as it overlaps with this megatrend with the eventual emergence of the energy information portal which will allow price signals, decisions, communications and network intelligence to flow back and forth in real time between suppliers and consumers. Customer engagement is a key to the future of the intelligent grid and AMI is the tool by which the customer is connected to the marketplace (see Chapter 7, The Destiny of Customer Engagement).

AMI will impact many parts of the organization as it provides the benefits of information, efficiencies and impacts changes in workforce and processes. Areas such as finance, customer information systems, engineering, asset management and work management will be impacted and need to be included in any process of strategic planning in order to take full advantage of the benefits (Figure 6-2).

The industry is finally starting to recognize the benefits of AMR/AMI to the overall system in order to develop the business cases necessary to garner the support of T&D management. This is certainly not always easy in an era when investments in the system, coupled with the high cost of natural gas are putting upward pressure on prices.

It is clear that AMR/AMI is at the core of the "grid of the future," bringing intelligence and a networked model to the electric, gas and water systems. Thus far, however, these end point devices are rarely viewed—or accounted for—at the level of the T&D system. Rather, projects are focused at the meter shop, billing system and customer interface level, leaving much of the potential revenue and savings unaccounted for in the models being used today.

The fact is that the IntelliGrid is not intelligent without AMI. Smart grid optimization is not "smart" without AMI.

As another layer to the AMR/AMI infrastructure, meter data management systems (MDMS) are the key ingredient here to gaining the most significant benefit from AMI/AMR as the repository and analytic tool kit that enables the use of the data—everything

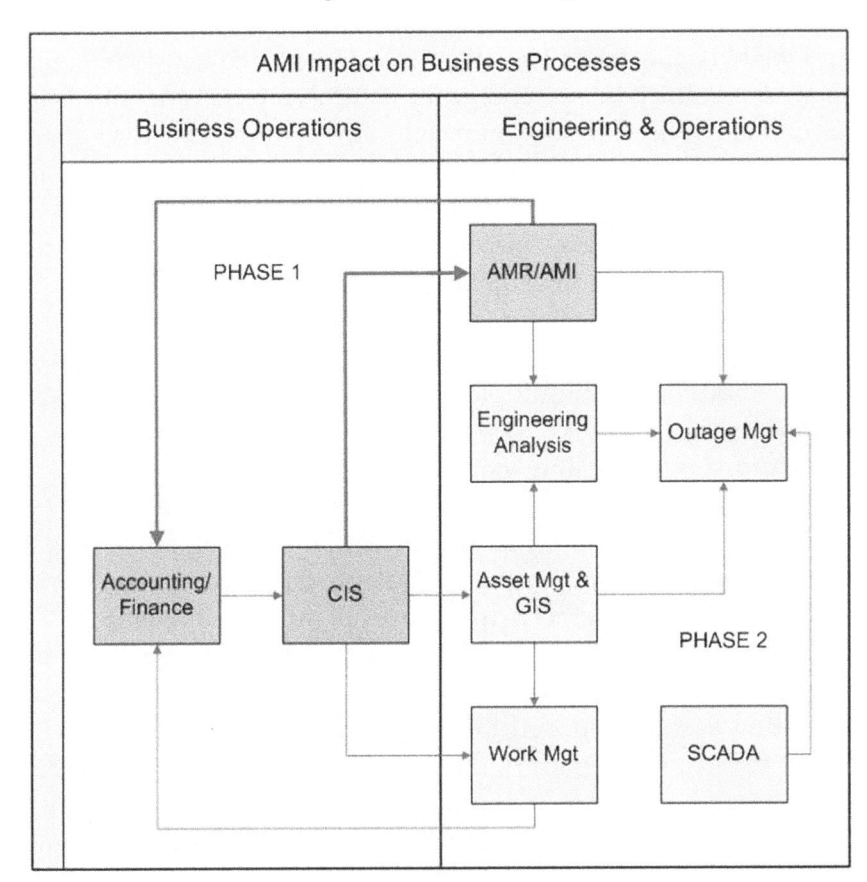

AMI Linkages to Business Organization

```
AMI Impact on Business Processes
┌─────────────────────────┬──────────────────────────┐
│   Business Operations   │  Engineering & Operations │
└─────────────────────────┴──────────────────────────┘
```

PHASE 1 → AMR/AMI → Engineering Analysis → Outage Mgt

Accounting/Finance → CIS → Asset Mgt & GIS

PHASE 2

Work Mgt   SCADA

Figure 6-2. **AMI will have impact across the organization as it affects and enhances operations in Finance, Customer Information Systems, Engineering, Asset Management and Work Management in addition to its own operational impacts.** *Source: R.W. Beck, Inc.*

from pricing and billing, CIS, load forecasting, demand response programs and customer segment offerings to extending load control research to drive energy efficiency and sophisticated pricing and rate programs.

AMI lies at the core of the smart grid, bringing clarity to the currently fuzzy picture beyond the transformers. Current SCADA (supervisory control and data acquisition) systems allow visibility to the transformer level and a modicum of control for reclosers and related equipment. However, in order to bring true intelligence to an operating system, end point control is required. All of this occurs across an information technology platform, such as Multi-Speak, which brings intelligence to the entire system.

Networks must have sufficient bandwidth to deliver relatively large quantities of data—meter readings may be as frequent as every five minutes. Control signals going from the utility to the meter will only add to that volume. Providers may provide services in addition to electricity such as gas and water as well as offerings for energy monitoring at multiple locations such as a second home or vacation cabin.

And down the line, smart appliances, plug-in vehicles and other smart technologies will be connected to communicate with the utility. Here again, the destiny of customer engagement plays a part through demand response programs, the consumer energy information portal and the virtual power plant (see Chapter 6).

A comprehensive AMI/AMR strategy provides substantial benefit to power providers including increased robustness in the operating system, reduced losses, lower costs and increased customer satisfaction. Typically, the largest financial benefit is attributed to the reduction or elimination of manual meter reading which can be as much as two thirds the cost savings. On the customer side, the benefits are impressive as well, including improved service quality and new service options such as demand response and time of use pricing. All in all, the whole package makes this a great option for companies interested in riding this megatrend.

AMI, MDM and AMR implementation are large investments for any utility and involve multi-year budgets. Companies must look at benefits and savings across the enterprise from billing, load research, capacitor control, improved outage restoration and customer satisfaction. The bottom line, however, is that this important technology piece of the intelligent infrastructure is and will be a key one for all utilities and power providers to succeed with this megatrend.

Distribution Automation

Distribution automation provides the remote monitoring of the distribution system and facilitates supervisory control of devices across the system—at the substation, feeder and customer levels. It also provides decision support tools to improve system performance and several application functions which help the operator to operate the system efficiently.

Power disturbances and anomalies represent a loss to the U.S. economy of 20 to 30% of U.S. electric system assets of $600 billion. While these losses are high, the opportunity for improvement is also great. The majority of all power outages and disturbances originate in the distribution network, therefore, the focus of achieving cost savings and improved customer service lies in automated, real-time response to adverse or unstable conditions on the system. Self-healing functionality of an intelligent system will instantly detect and react to power disturbances, improving operational efficiencies never before possible and with minimal customer impact.

As the intelligent grid evolves, the demarcation between distribution management systems and outage management systems will begin to disappear. These separate functions will combine and expand to handle additional capabilities such as SCADA and telemetry including customer calls. With real-time performance across all aspects of the system, a utility could stay on top of any worst-case scenario such as a large storm or hurricane with optimal operational efficiency of system operations, outages, downed lines, and safety issues.

Under pinning the intelligent distribution system is the knowledge and intelligence derived from system data and analytics. This key piece of the intelligent grid provides the high-level presentation and information and a bridge is built between IT and operations. This process requires the integration of many data sources throughout all system operation, is the hub of what essentially makes smart distribution operations smarter and key to optimization of the entire energy enterprise.

Distribution utilities are under pressure to meet the needs of a digital world, yet investment in smart distribution systems continues to be a challenge. By some estimates, distribution information and communications systems are installed at less than 75% of North American electricity substations and distribution automation penetration at the system feeder level is estimated at only around 15-20 %. In addition, more than 2.5 million miles of wire in North American distribution systems have absolutely no data collection or communications backbone associated with them. At the same time, the distribution network is facing massive upgrades as over one million miles of distribution wires installed over 40 years ago will soon need to be replaced.

Yet it is through the intelligent grid that the successful utility, taking advantage of this megatrend, will be able to meet the new business objectives of the 21st century. It is no surprise that many will present a strong case that a fundamental investment in a detailed distribution strategy and technology road map should be at or near the top of a utility's to-do list.

FACTS: Foundation for the Smart Power Grid

Equipped with 1950s-vintage electromechanical controllers, today's power system gives electrons free rein to flow according to the laws of physics, creating congestion, loop flows, and bottlenecks that undermine the efficient delivery of power from producer to consumer. Further, because the aging system often can't prevent voltage disturbances fast enough to prevent propagation, operators may restrict line throughput to well below capacity limits to preserve system stability.

Flexible AC transmission system (FACTS) technology (Figure 6-3) addresses both problems. The family of solid-state high-voltage electronic controllers comprising FACTS, developed by EPRI and Westinghouse, directs electricity flows with split-second precision, allowing system operators to dispatch bulk transactions smoothly along designated paths. And because FACTS devices respond almost instantly to disturbances, they help maintain system stability and reliability while increasing line capacity by up to 50%.

Further, FACTS and even more advanced controllers will pave the way for the smart, self-correcting digital power system of the

Flexible AC Transmission System: FACTS

Figure 6-3. A high-speed, electronic device using FACTS technology affords unprecedented control of the flow of electricity on high-voltage transmission systems operating on the New York power grid, providing a glimpse of how solid-state, semiconductor switches can relieve bottlenecks and revolutionize power networks in the future.

future that can help power companies work with this megatrend and not against it. Integrating energy and communications to support a fully functioning retail energy marketplace, such controls could usher in a revolution in consumer services.

Sensors: Enabling a Self-correcting Grid

The smart power grid will have self-monitoring and self-correcting features, enabling it to anticipate and respond automatically to disturbances while continually optimizing its own performance. To this end, the grid will incorporate an array of intelligent sensors, software and communications devices integrated with power system control functions. These technologies would complement FACTS devices to increase the control, capacity, security and reliability of the grid.

Dynamic thermal circuit rating (DTCR) technology lets system operators know the actual operating condition of a circuit in real time. Currently, engineers make conservative calculations of the load capacity of a line under any set of weather conditions and the line is operated at that rating. DTCR sense actual local temperature, solar radiation and wind speed and factors those calculations into a line's rating so that load could be increased or decreased appropriately. In some cases, the line can be operated at much greater capacity. Physics limits how transmission circuits can be loaded. In the eastern US circuits are underutilized by 10-15% and in the west by 20-30%. Using advanced technologies like FACTS and DTCR, the utilization of those lines could be greatly enhanced.

Fault anticipators could be used to increase security and reliability of the system. This technology allows very small electrical signatures on a power system to be monitored which in turn could conceivably prevent major power outages such as that of August 1996. In that outage, millions of customers in the western states were left without power when an overgrown tree in Spokane, Washington, contacted a line. It took an hour and twenty minutes to determine what had occurred. In such events, there is usually minor arcing before a major outage is triggered. With fault anticipators, the sensor would send information to the system operator, alerting them of the condition and its location, so that a crew could be dispatched to fix it before it became a problem or endangered anyone, so the people served by the line are unaffected. These technologies and others will form a self-correcting system that would be carried all the way into the substation level, so that even a substation could repair itself and notify the operator.

FACTS technology demonstrates the impact that these technologies make possible. FACTS devices move power on the grid like routers move messages on the internet. Operating with the speed and precision of microprocessors, these devices permit more electricity to be carried on existing transmission lines and react almost instantaneously to disturbances. Because they can act quickly enough to provide real-time control over the power system, FACTS devices can increase or decrease power flow on particular lines to alleviate system congestion. In addition, these controllers can enhance system reliability by counteracting transient disturbances almost instantaneously, allowing transmission lines to be loaded closer to their thermal limits, increasing capacity.

The unprecedented control made possible by power electronics presents strategic business advantages to early adopters. For example, digital control of the power grid would enable companies operating transmission systems to turn today's congestion problems into financial opportunities, delivering more power to customers and increasing revenues. Consider one of the infrastructure issues underlying the California power crisis. During periods of peak demand in the summer of 2001, Path 15, a notorious 90-mile transmission bottleneck in the middle of the state (Figure 6-4), could not deliver enough megawatts from southern California to prevent rolling blackouts in the capacity-constrained north. As a result, northern California power prices skyrocketed and the ISO was forced to impose rolling blackouts. With enhanced power flow control, however, the needed megawatts could be routed safely through the bottleneck to end-users.

The Western Area Power Administration completed the upgrade of Path 15 by adding a new line that increases the transmission capacity between Northern and Southern California by 1500 MW. This upgrade may relieve the bottleneck, but it does so using an old approach and old technology—underscoring the lack of incentive for investing in transformative new transmission technology.

Increasing line capacity with FACTS is a cost-effective, environmentally viable alternative to building new transmission lines—and more line capacity means companies can move and sell more power. And an added benefit to companies riding this megatrend on integration and delivery, for their part, customers benefit from enhanced reliability, better power quality, and potentially lower power prices.

Superconductivity: Breaking the Bottleneck

The technology with perhaps the greatest potential for revolutionizing power delivery is superconductivity. Operating with near

California's Path 15

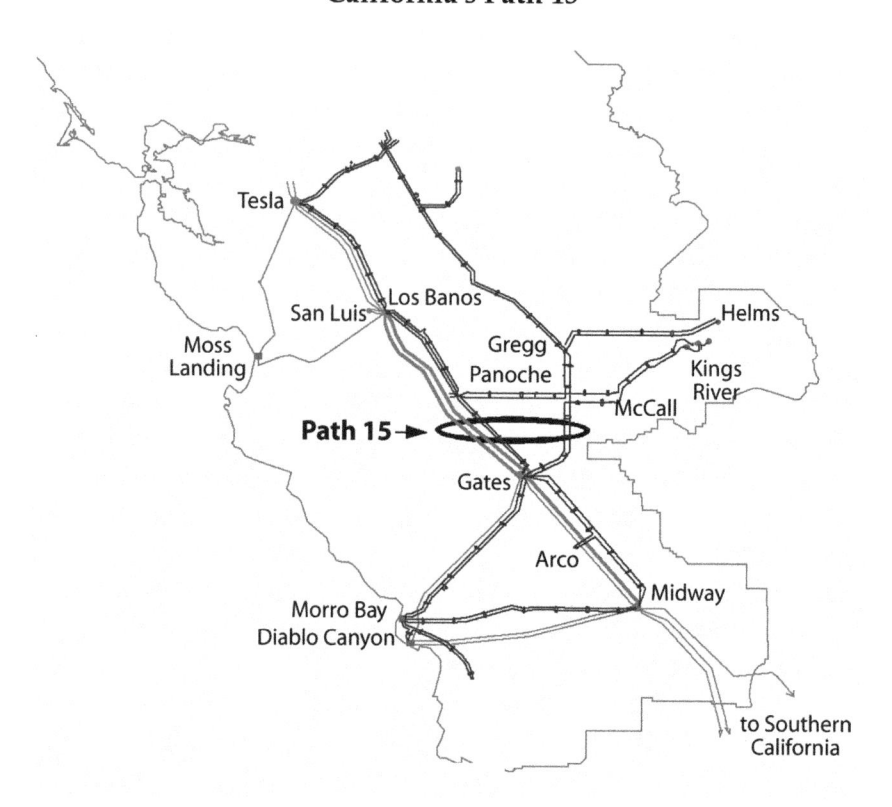

Figure 6-4. California's infamous Path 15 represents a classic example of the bottlenecks that can exist in the nation's current power grid. During periods of peak demand, these bottlenecks hamper efforts to prevent rolling blackouts. Path 15 was recently upgraded but not with the most advanced technology available.

zero resistance, high temperature superconducting (HTS) wire can carry three to five times more current than conventional copper and aluminum wire. The superior power density of HTS wire could enable a new generation of power industry components such as cables, transformers, fault current limiters and superconducting magnetic energy storage (SMES).

With HTS technology, electric companies, eager for solutions that let them deliver more electricity without constructing new conduits, could meet demand growth in urban areas while stimulating economic development—in turn creating more customers and increased electricity sales.

Discovered in 1911, superconductors are materials that carry electricity with very low or zero resistance loss. However, the need to cool such materials to extremely low temperatures with liquid helium precluded their use for decades. The situation changed for the better in the late 1980s, when researchers discovered ceramic-based HTS that can be cooled with liquid nitrogen—a lower cost process than with liquid helium.

Superconducting Cable

Figure 6-5. Detroit Edison's Frisbie substation is the site of the first installation and demonstration of an underground HTS cable in a U.S. utility network. Investment in HTS cable—and superconducting equipment such as transformers and fault current limiters—promises strategic competitive advantage that extends far into the future.

HTS cable could transform power delivery. On long-distance transmission circuits, HTS cables could provide a lower-voltage parallel path, relieving high-voltage transmission constraints and improving the capacity, efficiency, and reliability of the nation's electricity infrastructure. And future breakthroughs offer potentials—such as the ability to economically transport remote renewable power to the grid—that have yet to be explored.

One of the highest value applications of HTS cable lies in repowering underground transmission lines, especially as replacements for overloaded cables in older urban areas where the ability to deliver more power through existing ducts could be the key to economic revitalization.

In a pioneering demonstration project, Detroit Edison installed the world's first HTS power cables in a utility network (Figure 6-5). At the Detroit site, three HTS cables were installed to carry 2400 A AC at 24 kV—three times the current carried by conventional copper cable. Installed in existing ducts, the HTS cables together replaced nine conventional cables over a circuit length of 400 feet. Shortly before the system was commissioned in 2003, project engineers discovered leaks in the cable's vacuum thermal insulation, which is essential to the system's thermal and dielectric performance. Although the Detroit Edison cable installation could not be operated as intended, the setback did not slow the market momentum for HTS technology in the power infrastructure. In fact, the project provided valuable lessons for a new round of demonstration projects.

Southwire Company in Carrollton, Georgia, dedicated the world's first high-temperature superconductor (HTS) power delivery system in 2000 to provide power for industrial use. The system, which includes three 30-m long HTS power distribution cables, provides electricity to three Southwire manufacturing plants. It is the first time a company has made the difficult transition from laboratory testing to a practical field application.

More recently, Long Island Power Authority (LIPA) began construction of the largest and highest-voltage superconductor electric transmission cable system to date. The cable system, when completed in 2007, will be 138,000 volts (138kV) and nearly one-half mile in length.

The project, being undertaken by a government-industry partnership including the U.S. DOE, will be the world's first superconductor cable installed in a live grid at transmission voltages and will carry more power than all previous high HTS cable demonstrations combined. The purpose of the project is to demonstrate the operation of an HTS cable within an electric utility transmission system.

Currently there is a two-year pilot project under way that will test a new second-generation HTS cable which will deliver more than 50 MW to more than 8,600 homes and businesses in suburban Columbus, Ohio. The joint effort between the U.S. DOE, American Electric Power and its suppliers, is using a new cable that is less expensive and easier to maintain than previous technology. The new cable, called a Triax cable, places all three phase conductors—one to carry each phase of a three-phase electrical transmission—concentrically around a single core. The cable can carry up to 3,000 amps of power, about three times that of other superconducting cable used in pilots under way.

As pilot projects continue to bring this technology closer to a commercial reality, investing in superconducting cable does more than help a company squeeze more power through a given circuit. This is a transformative technology that can bring potent strategic leverage and competitive advantage to companies that take a systems approach to exploiting its capabilities.

This will be demonstrated in many ways. By enhancing delivery efficiency and reducing line losses, super-conducting cable allows more electricity to reach end users and perform useful work, with no increase in CO_2, NO_x and other emissions from fossil fuel power generation. That can mean a reduced need to generate or purchase power to meet a given level of customer demand. It can also translate into emission credits and offsets. Moreover, HTS may provide a needed boost to renewable generation such as wind, solar, biomass and geothermal. Some of these resources can't be exploited due to grid capacity constraints or because of long-distance line losses. High capacity, low-loss HTS cables could accept this clean renewable power and deliver it to population centers.

Looking Forward, Thinking Big

Investment in superconductivity also paves the way to increased use of hydrogen as an energy carrier and toward the futuristic visions of a continental supergrid and supercity energized by a symbiosis of superconductivity, hydrogen, and nuclear technologies.

The supergrid, an ambitious concept envisioned by the late EPRI founder and president emeritus, Chauncey Starr, would be a high-efficiency, coast-to-coast underground energy corridor for transporting both electricity and hydrogen. The supergrid would use a high-capacity (40-80 GW) superconducting power transmission cable cooled with liquid hydrogen produced by advanced nuclear power plants and pumped through the cable's center. The cable would cross North America in a giant loop to support the future hydrogen/electricity energy economy. An even more advanced concept would add transportation to the underground corridor in the form of high-speed electric trains. Power electronic converters would connect the DC supergrid to high-voltage AC transmission substations.

The supergrid concept simultaneously addresses key energy and environmental challenges by enabling a transition to sustainable energy resources and the efficient transport of energy in an environmentally benign manner.

A related concept, the supercity, envisions a future metropolis supplied with baseline power from advanced gas-cooled nuclear reactors and supplemented with solar and biofuel power as well as energy storage using hydrogen. Energy would be delivered through an electricity/hydrogen superconducting cable similar to that envisioned in the supergrid concept.

Supercity substations would distribute electricity on a local scale as well as generate and store hydrogen using reversible fuel cells. Surplus power from rooftop solar and waste-biomass sources would be converted to hydrogen at the substations. Transportation would be based on electricity and hydrogen, with electrically driven underground rail systems and hydrogen powered trucks and buses. Personal vehicles would employ hybrid battery-hydrogen propulsion.

Developing the supergrid and supercity would be daunting engineering challenges comparable to the building of the transcontinental railroad, Panama Canal, and the space program. We present them here as examples of the kind of bold, "out of the box" thinking needed to address the problems of energy supply and environmental constraints associated with continued population growth over the next century.

Succeeding with the Destiny of Intelligent Infrastructure

Investing in new infrastructure at a time of rising energy prices and rising demand remains the major challenge for utilities. Trying to balance the needs of a vastly improved and intelligent system with the cost requires a change in management philosophy, risk taking and regulatory mindsets. The fear of technology failure is not really at the core of this issue. It is rather the fear of making an investment with an uncertain time horizon and payback. Certainly, there is some technology risk in areas such as superconductivity, but adding sensors and control across a broad part of a system represents costs without significant risk. The challenge is that the costs cannot easily be justified in a short period of time. The good news is that the destiny of the capacity/carbon conflict will place a premium on efficiency and demand response as a way to ease the burden of building new power plants. This may put pressure on intelligent grid investment and commitments to customers to deliver the capability to secure these types of services. The challenge will be living up to that promise while balancing cost.

The existing electricity grid is inadequate to meeting the demands of the 21st century and must be modernized. We have heard radicalized statements that the grid is "third world" which is an exaggeration as in its current state continues to support the world's most vibrant economy with a remarkable (an enviable) record of performance. Nevertheless, the opportunity in the market is to expand, upgrade and build intelligence and communications into grid is the key component in reacting to all of the megatrends. If there is a single rallying point though which the megatrends intersect it is in the destiny of intelligent infrastructure.

The smart power system of the future will incorporate digital controllers and sensors to integrate energy delivery and communications, optimizing the overall performance and resilience of the system. The new "self-correcting" infrastructure will continuously monitor and correct itself to ensure reliable flows of high quality power.

The opportunities (and needed investments) are great but just as with the current system not everything is constructed at once.

Where to Go from Here

While we have provided examples of specific opportunities which can target the needs outlined in the three areas of this megatrend, strategic planning which incorporates several of these investments within the structure of competitive regulation, can lead to significant business growth and protect the market space available to these utilities.

- Implementation of advanced metering, distribution automation and a build out of the infrastructure to achieve digital control of the grid allows most of the destinies to be managed and creates opportunities for all stakeholders. Business cases today are often constructed using yesterday's paradigms and can fail to recognize the significant changes that will result in processes, engineering and system management. While it is difficult to envision the future changes of the system based on where it is today, the opportunities will come for those companies who make investments in the space recognizing the benefits will come in the future.

- Advanced Transmission technologies will allow increased efficiencies so that a variety of generation resources—including renewables—can be deployed. Substation automation and distribution management systems will allow greater reliability and availability as parts of the country suffer from capacity limitations and can provide the ability to reduce the reliance on a declining and aging workforce.

- AMI will bring a new level of system operations and customer interaction. The valuable data generated by the systems will expand utility visibility into its workings—and will interlink customers in ways that have never been seen. Whether it is remote disconnection transformer sizing, appliance control and management, AMI can have a profound effect on both sides of the meter. The added intelligence will also require new ways of managing—and opportunities for additional service offerings. These may not come through traditional channels. Social aggregators (i.e. church groups, communities etc.) may, for example, use AMI as a tool for reaching out to constituencies. Similarly, traditional utilities of all types will now have powerful communications tools and sales vehicles.

- New business models and opportunity will result, including the supplying of higher quality/reliability of power or a host of services for end consumers. Retailers may be able to tap into consumer demands, such as pre-emptive service calls on appliances before an emergency. Utility vendors such as tree-trimming services might be called out for tree-line contacts prior to failure, automatically based on fault indicative sensing.

The infrastructure investment model can help companies make sound investment decisions that balance risk and return while meeting the electricity needs of a robust digital economy

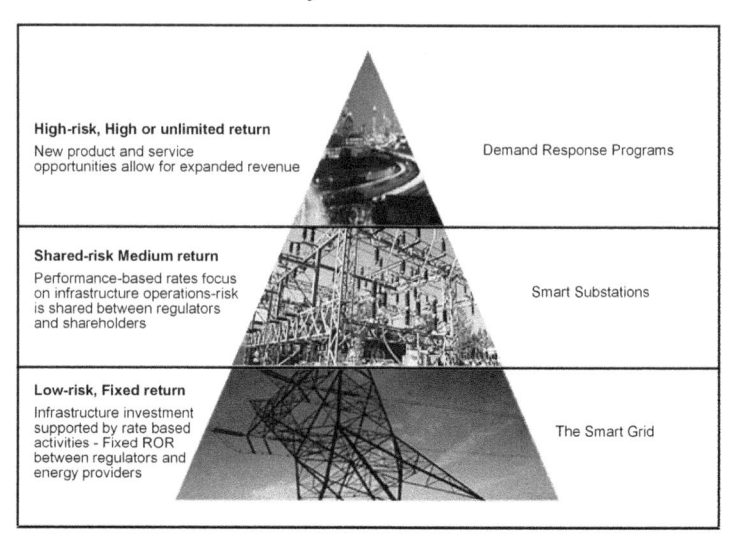

Power Delivery Investment Model

High-risk, High or unlimited return
New product and service opportunities allow for expanded revenue — Demand Response Programs

Shared-risk Medium return
Performance-based rates focus on infrastructure operations-risk is shared between regulators and shareholders — Smart Substations

Low-risk, Fixed return
Infrastructure investment supported by rate based activities - Fixed ROR between regulators and energy providers — The Smart Grid

Figure 6-6. Rate-based activities would include investment in transmission and distribution infrastructure technology such as digitally controlled FACTS. Performance-based investment may include smart substations while a higher-risk activity would consist of new service opportunities such as demand-control.

(Figure 6-6). This balance provides the power provider with an incentive to invest in technology to deal with the destiny of intelligent infrastructure, while offering an upside to those other firms that want to take a risk in using technology and building a new business model.

The framework for the 21st century power system is the smart, digitally controlled grid that replaces today's slow electro-mechanical switches with real-time power electronic controls. Investment in digital control (such as FACTS controllers and related technologies) should be supported by rate-based activities on a fixed-return basis.

The next level of performance based investment builds on the digital framework with technologies that integrate communications into the grid to support the real-time exchange of both information and power. This is the capability need to enable retail energy markets, power interactive service networks and provide self-monitoring and self-correcting features that instantaneously sense and counter disturbances. Infrastructure investment examples include smart substations that incorporate self-diagnostics and fault anticipation technology to improve infrastructure operations and optimize maintenance. This level of investment brings medium risk and return, shared between regulators and shareholders for IOUs, owners for co-ops and communities for municipal utilities.

At the top tier of the investment model are technologies that enable new service opportunities and expanded revenues. These would include devices to ensure reliable interconnection between the grid and distributed energy resources, and interfaces with building energy management systems to enable dynamic demand response.

References

"Cost of Power Disturbances to Industrial and Digital Economy Companies," Primen Market Study completed for E2I/EPRI CEIDs Initiative, (Consortium for Electric Infrastructure to Support a Digital Society), June, 2001.

"An Analysis of the Consequences of the August 14th 2003 Power Outage and its Potential Impact on Business Strategy and Local Public Policy," Mirifex and the Center for Regional Economic Issues, February 2004

"Electricity Sector Framework for the Future, Volume I," EPRI Report August, 2003

"Smart Power Delivery: A Vision for the Future," *EPRI Journal*, June 2003

"EPRI Study Shows Electricity System Improvements Needed To Prevent Economic Losses in 21st Century Digital Economy," EPRI news release, July 16, 2001

"Smart Power Delivery: A Vision for the Future," *EPRI Journal*, June 2003

Grant, Paul. "Energy for the City of the Future," *The Industrial Physicist*, Vol. 8, No. 1 (March/April 2002), pp. 22-25.

Silberman, Steve. "The Energy Web," *Wired*, July 2001, pp. 114-127.

Insull, by Forrest McDonald. 1962, University of Chicago Press.

NERC "2006 Long-term Reliability Assessment."
"Meeting U.S. Transmission Needs," EEI, July 2005
Salvatore Salamone, "Superconducting Sizzle," *EnergyBiz Magazine*, 11-12/06
"San Diego Smart Grid Study, The Energy Policy Initiatives Center, USD School of Law," prepared by SAIC Smart Grid Team, 10/06
"Meeting U.S. Transmission Needs," EEI, 7/05, prepared by Energy Security Analysis, Inc.
"Creating a Smart Distribution System One Process at a Time," Advanced Control Systems, 12/06
"Expanding the grid," *Energy Biz*, 3/30/07

Chapter 7

The Destiny of Customer Engagement
A Megatrend of Choice

*I*nsull struggled to find the best approach for measuring and charging for serving electricity customers. On a trip to England in 1894, he discovered the Wright demand meter and used it in Chicago to discount non-peak rates and further encourage consumption. For most customers, the ubiquitous electric meter and monthly bill have remained the most familiar and unchanging symbols of their relationship with their electricity provider.

I am convinced that if the rate of change inside an institution is less than the rate of change outside, the end is in sight
Jack Welch, former CEO, General Electric

Many industries have undergone dramatic evolution in which the customer became engaged at a much deeper level than ever before. This change has resulted in raised expectations of service on behalf of the customer in terms of value, turn-around time and company knowledge of an individual customer's situation. It can also result in the demands of the marketplace outpacing the industry's ability to deliver on those demands—and turmoil may be the result.

The internet has completely transformed the customer's involvement in shaping markets, giving the never-satisfied consumer new power over how companies service their customers. Whether it is McDonald's offering of "Have it Your Way" or the customized selection of music through iTunes, consumers are becoming conditioned to increased flexibility, better pricing and the power to decide the what, when and where of their buying decisions—such as "zero percent financing" or "$1,000 cash back," "nights and weekends are free" and that it takes two to drive in the commuter lane.

Besides the societal "training" of consumers to have more choice in everything, the technology of electric power demand has changed as well. The growing need for dramatically increased power quality, reliability, availability and security means that consumers now have back up batteries in their home offices and businesses with sensitive loads such as hospitals and semiconductor manufacturers find it necessary to install extensive back up systems—a service that is not provided by their power provider. More and more customers will seek out and deploy distributed resources for financial, security and environmental reasons. The need for consumer flexibility and choice will evolve further into more consumer programs such as demand response, time of use rates, green energy options, energy management systems and advanced, seasonal energy purchases and pre-payment of energy. All these options will take hold in bigger and broader venues.

Customers can already opt for fixed prices and flat billing in many jurisdictions. The concept of buying forward—such as prepaying for minutes on cell phones—will spread to energy. The practice of negotiating better rates will grow for businesses as well, both in industrial and commercial sectors and eventually residential markets. Large companies will negotiate pre-payments for discounts that will allow suppliers to get revenues in advance in exchange for the certainty of sale.

Customers will engage in markets in new and unusual ways, creating opportunities as well as challenges for power providers. Customer choice in electricity will not require formal deregulation—the clients will deregulate themselves, disengaging from and driving market forces via energy efficiency, demand energy programs and renewable energy resources.

As technology allows, this megatrend represents the impact of customer engagement upon the face of the electricity energy business. Customers are growing accustomed to choice and will demand their energy suppliers be flexible in their offerings, billing options and products. This does not necessarily mean choice *of* supplier but it does mean choice *from* their supplier, whether it is time of use rates, levels of quality and service or changing billing cycles to meet their pay schedules.

This chapter looks at the challenges facing the end-use side of the electricity enterprise, and opportunities for reinventing the utility/customer relationship. The megatrend of choice, borne by the technology of power demand, will require successful power providers to meet consumer demand and enable and promote a new paradigm of customer empowerment.

It's Still Your Father's Energy Interface

Today the meter remains the focal point of customer interaction. It also sits squarely in the path of the future by preventing needed changes in customer engagement with the electricity system. Just as the analog telephone system prevented a host of new technological service and features and was the main block to freedom in the marketplace, so too is the basic meter that strands consumers on one side of a technological road and the electricity suppliers on the other. This vestige of 1894 keeps markets from truly functioning, features from being added and disconnects the use of the product from its purchase.

Samuel Insull and Thomas Edison understood that their business was not selling electricity, but light—and it was soon apparent that they were also providing their customers with a variety of services, including warmth, comfort, safety, entertainment, and the ability to work more productively and efficiently. But despite the capacity of electrification to improve life and enable progress, power providers and customers alike have continued to perceive electricity not as a service, but as a commodity product and an entitlement.

Many utilities have labored to break out of this traditional mindset and offer value-added energy services such as energy efficiency programs and power quality consulting. Such services were supposed to flourish under the opportunities deregulation was expected to provide. However, efforts to protect customers—such as California's scheme to open wholesale markets while continuing to cap retail prices—have served to block progress toward an efficient open market where customers can choose from an expanded selection of energy services and providers.

Partial deregulation perpetuates a basic problem: At the retail customer level, demand for electricity is disconnected from the varying costs of supplying electricity. Until consumers become full partners in the electricity marketplace, efforts to protect them will continue to foster the notion of entitlement, limit their ability to make choices, and distort marketplace dynamics and the efficient use of energy and capital resources. Energy efficiency, instead of being a natural consequence of price, has to be imposed and subsidized at a cost to the public. After selling electricity at fixed prices for 100 years, the move to an open market where customers are exposed to prices moving in both directions has proven to be unpalatable. The fallacy of deregulation was the assumption that prices would only move in one direction—downward—which is unrealistic given market conditions as well as current technological limitations.

For this situation to change, customers will need exposure to price signals and the ability to make and execute decisions about their energy use. In this context the traditional meter is a dysfunctional relic of the mid-20th century commodity mindset. The old meter doesn't give customers—or utilities—meaningful information that they can act upon, and thus stands as a barrier to greater customer participation in the marketplace. Certainly, some customers will not want to be engaged in the market and for them aggregators will serve the function of providing price stability—of course at a cost. For others, the ability to control their demand will be attractive as will the risk of seeking power on the open market.

Five years after the nation's largest power crisis, the prevalence of advanced metering and dynamic pricing remains in the single digits. The Energy Policy Act of 2005 requires companies to offer customers time-of-use rates. Even so, ironically, though the customer programs possible through today's advanced meter can help companies overcome price volatility, extreme peak loads and poor reliability and offer customers cost relief, they still are approached with hesitancy by many companies.

State regulators have also held back on implementation of variable pricing options for customers by not requiring this as a part of service in restructured states. This has been the case, most likely out of a desire to protect the consumer from having to make complicated choices and which serves as a protective mechanism against price fluctuation. Yet, these very choices can mean lower prices to the customer as well as system stability and reduced operating and capital costs to the utility. The transition of electricity use from default electric rates to time-varied rates seems to also be a challenge even though customers are used to varied rates for services such as phone service.

Yet, the incentives are there for power providers without looking too hard. A FERC-financed study states that a moderate amount of demand response could save about $7.5 billion annually by 2010. And a Rand Corporation study came to a similar conclusion saying that the utility industry overall could save between $50 billion and $100 Billion over the next two decades if demand response becomes the norm.

The meter is the last vestige of the old industry acting as an "iron curtain," blocking system intelligence from flowing to and from the user. This protective layer provides comfort but little else, much as cash registers of old versus today's point-of-sale systems. The advent of point-of-sale instantly linking purchases to inventory and production revolutionized retailing, distribution and manufacturing. Removal of the meter and its replacement with advanced technology will do the same for the electricity business.

For most of the 20th century, the electricity sector was defined by the technology of supply: ever-larger power plants produced cheaper power and an expanding grid delivered that power to consumers. In the 21st century, however, the electricity sector will be increasingly defined by the *technology of power demand*—specifically, intelligent customer-based devices that enable ever-broader consumer participation in electricity markets, and access to an array of new services based on the convergence of energy and communications.

The pathway to a new, vibrant retail electricity sector lies not in protective regulation, but in *customer empowerment*—the transition to this new business model is happening now, however slowly as the destiny of customer engagement. Forward looking power providers will see the value in giving customers the information and the tools to play an active role in electricity markets, the information and tools that they are beginning to demand at an increasing rate and the power to create benefits not only for themselves, but for the electricity provider as well.

Tapping Technologies of Demand: The Intelligent Energy Portal

The intelligent grid will unlock opportunities for a new two-way relationship between customers and energy companies—especially when the smart grid's advantages are brought into customer buildings via an intelligent energy/communications portal. (Figure 7-1) The disappearing consumer—singularly neglected under deregulation—could thrive again with the introduction of the energy portal. The portal essentially replaces the traditional electric meter with a two-way interface integrated with the customer's building energy management system (EMS). EMS, acting on pre-programmed instructions, would free the consumer from day-to-day decision making, controlling the operation of the electricity-using equipment in the building, from space-conditioning systems and lighting to computers and appliances. Even individual appliances could become smart. Air conditioners and refrigerators with embedded "smart" sensors could adjust operations according to price signals—as well as monitor themselves and even arrange for their own replacement or repair.

Real-time pricing and profitable demand management are just

the beginning. The portal will also make it easier to connect to distributed resources, enhance lighting and temperatures control, and optimize industrial systems. And still to come are an array of new energy-related services centered around power quality monitoring, green power, security, entertainment, and more.

The Intelligent Energy Portal

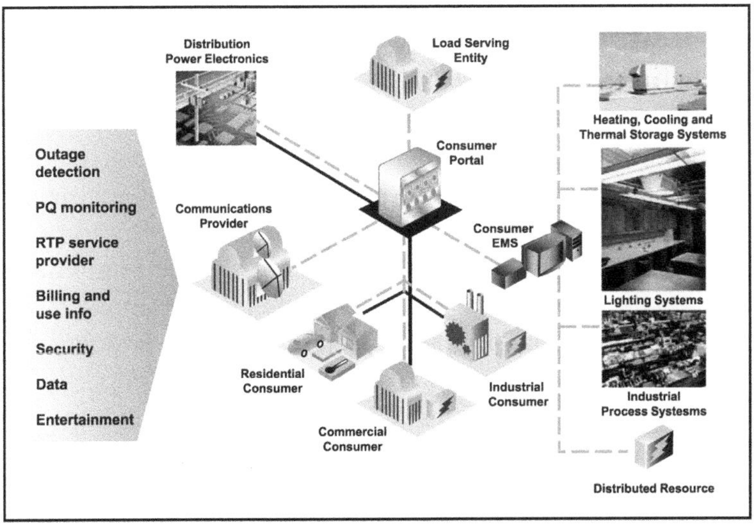

Figure 7-1. The energy portal is key enabler of the smart power delivery system, providing customers with the ability to reduce consumption and costs and take advantage of a new generation of services such as real-time pricing and enhanced energy/information management as diverse as those offered in today's telecommunications market. *Source: "Power Delivery System and Markets for the Future," EPRI Technical Report 1009102, 8/03.*

The development and widespread implementation of the consumer energy portal is an essential step toward a new customer service paradigm in which customer empowerment works as a tool to combat protective regulation. In parallel, the regulatory paradigm changes from protecting consumers to *protecting markets* and unleashing innovation.

Perhaps the most important capability the portal will enable is real-time demand response. For free markets to function, consumers must be able to receive price signals and respond to them by adjusting their consumption—the basic principle of supply and demand. In regulated markets, customers can respond to price changes over long periods of time, such as months or years. However, they generally can't respond to the rapid price fluctuations of real-time bulk power markets, for both technical and institutional reasons. This situation leads to bad consequences, as we saw in California, including over-consumption during peak periods, price volatility, rolling blackouts, and the need for new peaking and reserve capacity. Still another consequence has been underinvestment in energy efficiency measures. If customers could see the actual impact of their consumption decisions on the price they pay for electricity, many would adopt more energy efficient equipment and practices without the need for regulatory intervention. By timing when they buy electricity, consumers would get better prices while reducing strain on the grid and improving efficiencies all the way up the electricity value chain. For all these reasons, customer participation is essential to a fully functional electricity market. It is important to note that this does not necessarily mean open access and a choice of suppliers. Most of the open market experiences have shown that customers care little about their given vendor but they care greatly about price and service. It is not necessary, therefore, for supplier choice; but it is necessary for service choice.

By enabling real-time demand response, the digital energy/communication portal will be the linchpin technology that ushers in that fully functioning marketplace. The portal will open the consumer gateway—now blocked by the old electromechanical watt-hour meter—allowing price signals, decisions, communications, and network intelligence flow back and forth between customers and energy companies. Of course, most consumers won't have the time or inclination to continuously monitor power prices and adjust their usage accordingly. The portal, operating in conjunction with an EMS in a home or office building, would perform most functions automatically once programmed by the homeowner or facilities manager.

A number of current, emerging technologies can be optimized through the portal and offered by the retail power provider as optional, or even free services. A prominent example of this is the provision of high-speed broadband internet access directly through power lines, or broadband over power lines (BPL), endorsed by the FCC several years ago. According to Chartwell's AMR survey of 100 utilities, 33% of electric utilities in 2003 reported using, planning or considering broadband services through their existing service. The catch is that hardware must be installed on-site to provide the power line carrier (PLC) and voice-over-internet-protocol (VOIP) technology necessary for BPL. Offered through the portal, this service can be provided using no additional hardware or technology at an equal or higher quality service than available today, and is particularly desirable in more rural areas that do not have access to cable or DSL technologies.

An example of the market leverage afforded by real-time demand response comes from a California experiment where a 2.5% reduction in demand led to a 24% reduction in price. In a program conducted for the California Energy Commission by Global Energy Partners a strategy was used to reduce power demand air conditioners, lights, motors and other electrical loads from commercial and industrial facilities throughout the state. The program utilized a combination of technologies that enable automated notification through web-based applications and telecommunication devices that activate both voluntary and automatic curtailments during peak demand periods.

This program is an example of a "killer app" in the new energy marketplace: profit-driven conservation (or market-driven demand response) in which both the power provider and the customer benefit from the ability to respond to price fluctuations in real time. This type of demand response represents the free-market reincarnation of the demand-side management (DSM) programs of the early 1990s, which aimed to reduce customer energy use during peak periods through energy efficiency measures and load control. Unlike the old DSM programs, however, this approach monetizes the energy savings of load management. The customers in this partnership don't just save energy, they make money by sharing the profits from the sale of power they've temporarily stopped using, at peak rates.

For energy companies, this approach offers another profitable twist: the capability to use multiple, linked customer buildings not only to reduce peak demand but as a dispatchable generation resource. In fact, the energy obtained through load curtailment could be bought and sold just as though it were the output from a power plant. This concept is demonstrated here as "the virtual power plant."

Profitable Energy Conservation: The Virtual Power Plant

Energy partner, power trader, conservationist. With the smart grid and advanced portal, words like these could replace the concept of "energy consumer." Just as the internet can transform the passive media consumer into a content creator, the smart grid and energy portal could change the customer side of the equation forever, making customers partners with power producers in ways that offer myriad benefits to both parties (Figure 7-2).

The multiple scenarios possible with these innovations have yet to be uncovered. In a straightforward example, these technologies could pass pricing and control information from the energy provider to a commercial customer. During periods of high electricity demand and prices, the customer's energy management system could dim lights, curb air conditioning, or reduce other non-essential power uses. Beyond saving customers money, these reductions would relieve strain on the power system, forestalling rolling blackouts. Ultimately, the energy saved could be bought, sold, and dispatched just like power generated at a plant.

Unlike earlier strategies to shrink peak consumption, this approach actually monetizes the energy savings of load management. Thus armed, customers will not only save energy, but share in the profits from the sale, at peak rates, of the power they chose not to use—a huge opportunity for the entrepreneurial power provider that should not be undervalued. Solutions such as this help the power provider reduce peak demand, manage capital costs, in-

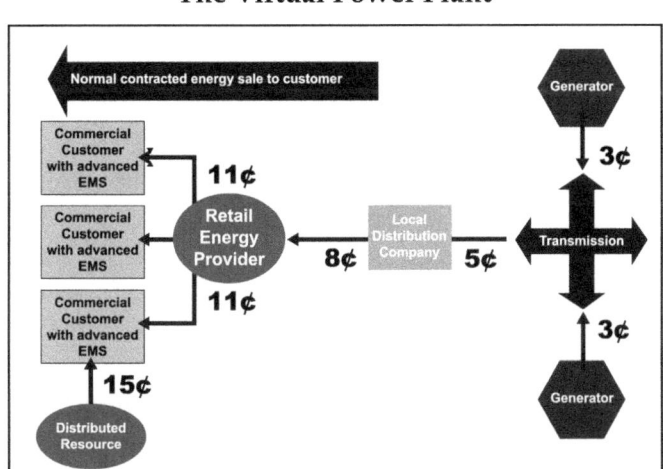

The Virtual Power Plant

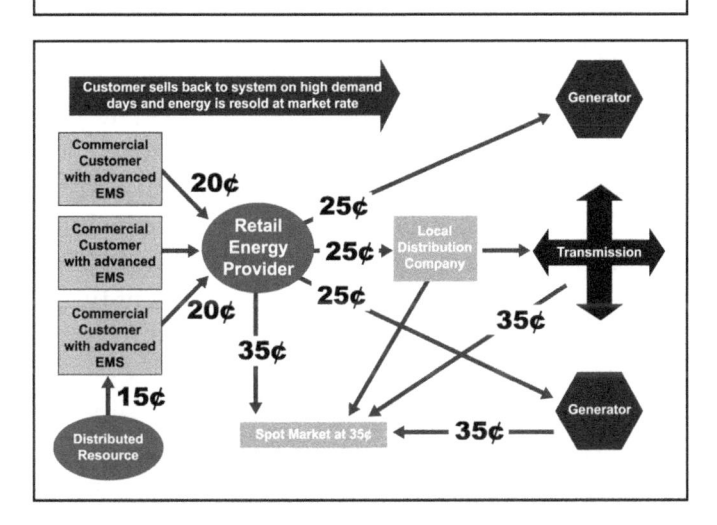

Figure 7-2. The smart grid and energy portal can provide a multitude of transformational scenarios such as the virtual power plant where pricing and control information from energy provider and customers can pass freely turning customer into seller of excess energy during high-peak times. In this scenario:
- During "normal" times, the retail energy provider purchases and provides electricity contracts on behalf of its commercial customers, acting as their agent.
- During high-demand times, the retail energy provider uses the energy management systems within buildings to lessen demand and resells power back to the grid for a profit, sharing that profit with its customers.

crease customer satisfaction and, ultimately succeed with this megatrend of choice.

Customer-driven Energy Services

The energy/information portal also opens the door to retail competition and differentiation of energy and communications services. For both consumers and providers, the portal provides a tool for moving beyond the commodity paradigm of 20th century electricity service and enabling a new set of energy/information services as diverse as those now offered in today's telecommunications market.

Skepticism among customers and service providers alike has clouded the long-term potential for retail competition. The general sentiment is that there are insufficient services to justify investment in intelligent metering much beyond automated meter reading and demand response, and there is no need for a better meter because the range of services is so limited. Breaking this chicken-and-egg cycle will be important to complete market transformation.

The potential for retail services is larger than currently imagined. In the future, for example, building on the capabilities of the energy/information portal, retail service differentiation could be taken inside the home or business, offering different types and levels of service to individual appliances with embedded microprocessors that communicate individually and directly to service providers. In principle, once inside the home or business, service opportunities could branch out dramatically.

An example is fuel switching for optimization of appliances such as air conditioners or furnaces. Residential, commercial and, particularly, industrial customers will want to make more active choices when it comes to fuels. There may be times when it makes the most sense to use electricity for HVAC units and at other times natural gas. Providers will need to be prepared for this as customers become more sophisticated in energy management options to optimize the system or lower cost.

Such expanded services would include true customer choice of energy provider and service—low cost, green power, high power quality/reliability—as well as a variety of value-added energy/communications services. One dimension of service could be enhanced energy efficiency based on smart end-use devices that communicate with the power system to reduce output during demand peaks, and that can self-diagnose and troubleshoot problems, even order replacement parts. Eventually the portal may also allow for cable telephone, TV, and broadband computer access.

There are companies today providing such services. Comverge and Enernoc recently went public with their business models of providing consumers and businesses alike with demand response capabilities that are sold back into the market at peak times. Site Controls, another player in this emerging industry, has created a network across the country for commercial and small commercial facilities that allows facility managers to control their environments as well as provide intelligent load management.

Would energy companies merge or develop partnerships with telecom/cable TV firms to deliver integrated power/communications/entertainment services? The real opportunity may be realized with the repeal of the Pubic Utilities Holding Company Act. Communications and entertainment firms may opt to become involved and acquire electric distribution entities. Even under the current business model, the marriage of technologies between communications and electricity through the wires solves the "last mile" challenge and brings a host of new business opportunities. The question may become not whether the electric providers could compete with existing broadband on a financial basis for monthly fees (they might not) but rather if the electric supplying entity could enter the market offering free broadband service in exchange for load control on peak days.

The portal would also enable an easy "plug-and-play" interconnection for on-site distributed generators, ensuring safe and reliable operation of the DER device and the grid. With a portal-enabled two-way interconnection, customers operating on-site power would have both independence and security. They would be able to produce some or all of their own power, with grid backup when necessary. They would also be able to sell surplus power back to the grid since the portal would serve as two-way meter, in this example running "backwards" to measure the power being sold rather than consumed.

Electric Transportation as Distributed Generation

In the future, the energy/information portal can help take the concept of customer-generated power to a new level: plug-in hybrid electric vehicles (PHEVs). Today's hybrid vehicles use an internal combustion engine for primary power and a relatively small electric motor for supplemental acceleration. Although these hybrids offer good fuel economy and reduced emissions, they are still powered entirely by fossil fuels. In a plug-in hybrid, an electric drive provides primary power and a small internal combustion engine provides supplemental power. Plug-in hybrids would recharge their batteries from the grid, providing a new load for power companies. Moreover, they could be an enabling technology and strategy for the successful introduction of hydrogen vehicles using fuel cell-battery plug-in electric power plants.

Development and commercialization of these new transportation technologies will require substantial R&D, but the investment could be shared through collaborative public/private partnerships that include the automotive and energy industries and government organizations. Austin Energy, the municipal power agency in Austin, Texas, has announced plans to use the stored energy in automobiles to supply the community at peak times.

Hybrid fuel-cell vehicles could serve as generators when they were not on the road (90% of the time for most passenger vehicles), helping to alleviate electricity shortages, providing grid regulation and spinning reserve, reducing pollution, and saving money for consumers. Upon arriving home, the consumer would plug the vehicle into the house to power home appliances, and even generate excess electricity to sell back to the grid. The portal's two-way metering function would run backwards to reduce the electricity bill. Such PHEVs could revolutionize the relationships between energy consumers and energy producers while meeting public policy objectives for improved air quality, dependable grid-level peak power production, and enhanced national energy security.

During off-peak hours, plug-in HEVs could charge from the grid with low-cost electricity, thereby improving power generation efficiency and reducing fossil fuel consumption. During peak hours, PHEVs would plug in and become distributed generators providing power to the grid. With time-of-day pricing, added value could be so significant as to convert PHEVs from an expense item for their owners to a valued capital asset with an attractive return on investment.

The Art of Energy Efficiency and Smart Control

The common thread throughout all these technology options made available by the energy portal is energy efficiency. In the face of concerns over climate change, rising energy costs and the outlook for increased electricity use by a proliferation of electrically driven devices, greater energy efficiency is the first choice and least cost solution. Increased energy efficiency makes the most of our energy resources, reduces the need to build costly power plants and minimizes the impacts of pollution.

A recent study by the Brattle Group examined the effects of reducing electricity use by a modest 3% during the hours of highest load demand for five Mid-Atlantic utilities. The results show that by reducing power demand at these times would reduce energy prices of non-curtailed load by at least $57 million to $182 million annually in this region. Depending on market conditions, this reduction can yield an average energy market price reduction of $8 to $25 per megawatt-hour, or 5 to 8% on average. The study also reveals that demand response participants would save about $9 million to $26 million for energy annually and another $73 million for capacity charges.

Like the previous example of demand response in California, this demonstration shows the dramatic effects that even a small reduction in energy use during peak demand can have when prices are high and how demand response programs and energy efficiency in general, implemented regionally can decrease the use of the most expensive sources of power, avoid power shortages and enhance reliability. The market impact in each individual zone would be substantially smaller if it curtailed its load in isolation. Other benefits that were not quantified in the study include competitiveness, price stability, deferred T&D costs and environmental implications.

The compact fluorescent light bulb is a good example of a an available technology that has offered a simple, solution for both residential and commercial applications saving a much as 50% on the energy used when combined with occupancy sensors or other EMS systems. Yet, appliances are transforming into devices with much more capability along with the inherent efficiency factors.

With the development of the intelligent grid and the energy information portal, advanced sensors are enabling a new approach to traditional energy efficiency and energy management down to the appliance. This capability will transform the customer role in energy use, allowing them to play an active role in solving power quality issues, improve productivity and efficiency, ease load control and optimize energy management and cost control.

With increased implementation of advanced metering and the energy portal, energy companies looking to succeed with the destiny of customer engagement will invest in the communications infrastructure and customer demand response programs and drive them to become more mainstream practices. The benefits can be huge for all stakeholders, customers and power providers alike. Yet, the next frontier for energy efficiency is in the end-use appliance itself.

Smart Appliances

Today, the electric power industry builds generation capacity to cover peak load with a buffer of 6% of expected demand. New smart appliances can help alleviate this buffer zone and take the edge off of stretched capacity on a hot day. Smart appliances can react almost instantaneously to power disruptions, giving power providers the time needed to put more power online. A few hundred thousand smart appliances in the system could reduce a utility's buffer by 1%.

Appliances such as refrigerators, water heaters and air conditioners make up about 20% of the nation's power load. This is a significant place to begin when approaching smart control to drive energy efficiency and reduced peak demand. Refrigerators will be able to keep themselves calibrated to the temperature necessary, lights will brighten and dim automatically in response to the available sunlight and to occupants in the room.

In the not so distant future, the country's current inventory of big-screen TVs, refrigerators, computers, compressors and equipment will give way to this new generation of smart appliances and machinery, connected and responsive to power availability as well as price signals that can, by some estimates, cut electricity consumption by as much as 25 to 45%.

When wrapped up into one efficient system, the combination of the energy portal, EMS and smart appliances has been called in some circles "the smart house." Keeping track of all energy use from the toaster to the television, customers can monitor how much energy is used and when. The smart house also lets people implement efficient security systems and safety features such as lights turning on when the alarm system goes off showing exits to the home. And in homes with elderly or disabled people, automated systems can help them maintain a more independent and active lifestyle. With services as simple as occupancy switches to turn on lights, changing the television channel or reminders for people with short-term memory problems, individuals can be independent longer. This issue interplays with the destiny of demographics as more and more people—baby boomers in particular—will be living longer and more independent, active lives. A company called Control 4 has developed a home interface which controls not only traditional entertainment appliances and lights, but links to advance metering infrastructure to signal the homeowner when an energy event is occurring and can automatically manage household consumption.

There are many opportunities and strategic benefits for utilities to invest in the development of these devices, how they communicate with the energy meter and affect the power system. The most basic are to offer incentives to customers why buy smart appliances, similar to the rebates offered today for energy efficient refrigerators and dish washers. But taking this on at a more strategic level can be a boon for the power provider concerned with staying on top of this megatrend of choice.

Succeeding with the Megatrend of Choice

Utilities of all types will be facing competition from new market entrants that provide a host of services desired by customers who are used to having choices. Other groups of customers will be aggregated by affinity groups such as churches, social clubs and neighborhoods with the goal of reducing prices while supporting causes. Finally, customers will demand control, levels of power quality and the ability to manage their bills and usage.

Clearly there is a significant expense in the core investment in the energy portal, and in establishing new service offerings based on new infrastructure. This is the case across the industry regardless whether it is a municipal, co-op or IOU. The challenge is in recognizing the investment will pay off over time in ways that cannot currently be anticipated in terms of added revenues, new services and lowered as well as avoided costs.

Until consumers become full partners in the electricity marketplace, the need to protect them will continue to foster the notion of entitlement and to distort marketplace dynamics, regulation and the efficient use of energy and capital resources. The megatrend of choice means a growing portfolio of diversified service offerings in addition to commodity sales will not only better serve the various needs of energy consumers, but will provide new lines of revenue for service providers. It is again worth stressing this is not about choice of suppliers—although it may be about the *type* of supply—it is about customer choices in how they engage with the market.

The real opportunity in the marketplace lies in increasing the functionality and value of electricity through consumer benefits. This functionality will far outweigh the cost and will allow the system to operate in ways never imagined. Of course, the real challenge is in the mindset of traditional utilities and models that have been built—and operated very successfully—for more than 100 years. This model has given consumers a choice of vanilla ice cream where the only differentiating factor was the price based on the quantity. It is little wonder that the real value of electricity which is enormous as the driver of the entire economy, is undervalued and under constant challenge. The mismatch of value to price is obvious as numerous other commodities such as minutes on cell phones, channels on cable television, songs on MP3 players and bottled water are purchased at double the price of gasoline requiring a greater and greater wallet share without raising consumer ire.

Opportunities lie in harnessing consumer power in terms of product purchased and managed. A classic example in the electricity business is in standby or small scale generation, pilloried as a bane on the system instead of incorporated as part of a distributed utility model. Demand response, the grown up sibling of demand side management, will allow customers to sell power back into the marketplace which can then be marked up and resold.

Just as IBM no longer makes PCs, Western Union no longer sends telegrams and Abercrombie and Fitch is no longer one of the nation's leading department stores, the energy business will be transformed because of customer engagement.

Where to Go from Here

While we have provided examples of specific end-use technology and service opportunities which can target the demands of the destiny of customer engagement, strategic planning which incorporates several of these investments within the structure of competitive regulation, can lead to significant business growth and protect the market space available to power providers. Some ideas for that include:

- It is important to positioning the utility as a leader in technology in order to take advantage of asset investments. No one is better positioned to make investments and bring advantage to customers. This positioning allows building a business for the future while reacting positively to the other destinies. Engaging customers requires intelligent infrastructure, leads to resolving the carbon/capacity conflict, helps ameliorate the demographic challenge while presenting new business opportunities.

- Creating a model of communications and investing in the right technologies for all constituencies will substantially improve energy efficiency and electricity intensity. This does not have to happen all at once but should be planned over a finite period of time.

- Involving the customer as a partner improves power system security and functionality. One of the side effects of this movement will be that rates and regulation will be driven more by customers than by traditional regulatory mechanisms. Self-selection into programs (such as paying premiums for green power or higher power quality) will mean a shift in cost-of-service calculations. Forced selection—such as remote disconnect for non-payment of bills—mean the overall costs for customers will go down.

- SQRA (security, quality, reliability and availability) will mean increased revenue opportunities and better system operations. This may be coupled with new financial models for all market participants.

- Automated metering infrastructure and distribution automation is the key to the marketplace and linkage to the end customer.

The infrastructure investment model can help companies make sound investment decisions and customer engagement strategies that balance risk and return while meeting the electricity needs of a robust digital economy and an increasingly sophisticated energy consumer (Figure 7-3). This balance provides the power provider with incentive to make the investments necessary while providing appropriate returns to those who want to reach into the market find opportunities within the destiny of customer engagement.

The development and widespread implementation of the energy/information portal will bring substantial benefits to both energy consumers and providers. The portal will provide the critical linkage that allows customer participation in the marketplace, reducing the need for protective regulation and increasing efficiency along the entire electricity value chain. The portal promises a payoff that is certainly worth the investment.

Energy Use Investment Model

High-risk, High or unlimited return
New product and service opportunities allow for expanded revenue — Value-added Energy Services

Shared-risk Medium return
Performance-based rates focus on infrastructure operations-risk is shared between regulators and shareholders — Load Shedding

Low-risk, Fixed return
Infrastructure investment supported by rate based activities - Fixed ROR between regulators and energy providers — Consumer Energy Portal

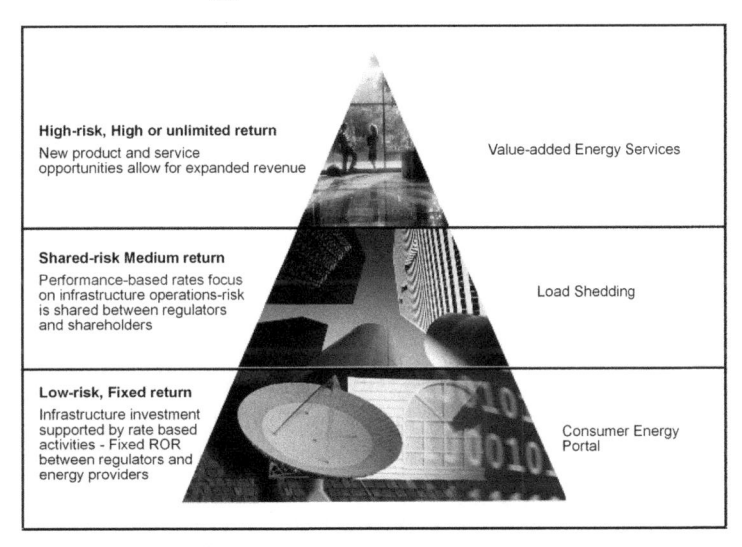

Figure 7-3. The advanced energy/information portal would be supported by rate-based investment and provide the foundation for a plethora of transformational product and service capabilities for all types of customers. Performance based investment may focus on load shedding programs and other activities to maximize productivity of the system and higher-risk investment activities may include a wide variety of value added services such as power quality, energy management and advanced pricing structures.

The investment may not be as daunting as it appears. Referring back to the infrastructure investment model (Figure 7-3), the initial investment in a portal (in the form of a smart meter) is an infrastructure investment supported by rate-based activities. The company then learns to use this meter to offload peak demand and resell power at a higher price, managing price volatility. The risk of developing these capabilities is shared among regulators, shareholders, and perhaps large customers. Once the infrastructure and service foundation is in place, a variety of new value-added service opportunities allow for expanded revenue.

Certainly since the buzz of deregulation began in the 1990s, the industry has sought its "call waiting" for electricity. It may be that the question was asked in reverse. That is, rather than looking for call waiting as an answer to increasing revenues, service levels and profitability, the industry should have been seeking infrastructure to support the new services. As discussed in chapters 2 and 3, telephony didn't decide on call waiting and then develop the infrastructure to support such a service, rather the telecommunications industry invested in infrastructure and technology that would allow for such services. By building the platform—in this case the portal—the industry will discover new potential business opportunities never imagined.

Resources

Insull, Forrest McDonald, University of Chicago Press, 1962
Electricity Sector Framework for the Future, Volume I, EPRI, August, 2003
"Power Delivery System and Markets for the Future," EPRI Technical Report 1009102, August 2003
Silberman, Steve, "The Energy Web," *Wired*, July 2001, pp. 114-127.
Ahmad Faruqui, "Breaking out of the Bubble—Using Demand Response to Mitigate Rate Shocks." *Public Utilities Fortnightly*, March 2007
Ken Silverstein, "The Time to Use Demand Response," *EnergyBiz Insider*, February 6, 2006
John Gartner, "Appliances Wipe Out Blackouts," *Wired*, June 22, 2005
"Turning on Energy Efficiency," *The EPRI Journal*, Summer 2006

Chapter 8

The Destination: A Transformed Electricity Sector

The energy business continues to evolve rapidly for all players in the sector, both public power and regulated markets. The challenge that all companies face is transitioning to a competitive landscape—whether that competition is for workers, power supply or customers. Even those markets which have backed away from open access find themselves in new territory as pressures to perform to new expectations rise, existing infrastructure requires revitalization and energy companies face a future of carbon constraints.

This future, however, is ripe with opportunity. Clearly, part of the challenge has its roots in history as well as in the past decade of instability. Confusion and uncertainty about the future has prevented many parts of the electricity sector from moving forward. In some cases, these issues have been used as a politically expedient excuse for not making decisions or as a rationale for cutting costs regardless of logic.

Many of the business decisions that have been made over the past 20 years have been reactive to short-term market trends as opposed to being proactive to take advantage of the longer-term and transformational megatrends. Companies reacted to the threat of deregulation by spinning off profitable portions of their businesses while trying to save stranded investments. Others became internally focused, cutting back everything in sight. Many IOUs made the lemmings' jump into foreign countries under the guise of "learning" about deregulated markets, only to find the painful cost of those lessons.

There have been some very solid winners as well. The nation's nuclear industry kept its focus and drove performance to unmatched lesions. Public Power used some of the confusing times to add fiber rings to their jurisdictions—investments that will pay off in the long run. Still others kept a rare focus on their customers and made investments in advanced metering and related IT infrastructure even without having the "perfect" answer.

The industry has gained some clarity with the settling down of trading markets in the wake of the Enron era, buoyed stock prices as well as the passage of EPACT 2005. The worry and risk of some investments has been ameliorated—but also replaced with other concerns such as finding capacity, weak infrastructure, siting transmission facilities and preparing for a new era of customer demand.

After examining the system stalemate and the possible consequences of continued inaction, this chapter will explore a vision of the future originally created by stakeholders throughout the industry and outlined in the 2003 Electric Power Research Institute (EPRI) publication, *Electricity Sector Framework for the Future*. Part of this vision looks to expanded intelligent infrastructure and is taking shape with significant investments in smart transmission and distribution as well as advanced metering. New generation technologies appear to be ready for increased scale, yet there is a very real state of fear amongst power providers when it comes to making the critical investment in these new sources of power, even with sorely needed capacity.

We then recall paths of transformation from the telecommunications industry, and reinforce the argument that such transformation occurs as the inevitable outcome of market demand for new products and services made possible by infrastructure investments—and not as the result of regulation or legislation. On the basis of these examples, we maintain that although some regulation may always be needed, any return to "regulation as usual" in the electricity enterprise—a real possibility should stakeholders seek easy solutions to today's quandaries—will fail to create a business environment that could give rise to the industry's vision of the future.

The Destination: A Transformed Electricity Sector

Finally, using the megatrends as the goal, we can envision the future in a way that offers sustainable electricity, rational pricing (since reasonable is a question of perception), engaged customers and expanded services.

Symptoms of Industry Malaise Continue
What are the symptoms of the current industry malaise?

- Many companies continue to operate as cheapest in class rather than "best in class." Though understandable, the consistent push to force costs out of the system has compromised the quality of the operation. Some firms are so driven by earnings before interest and taxes (EBIT) that they have forsaken the industry's valuable legacy of running the world's best energy provisioning system.

- The regulators' decision to freeze prices (supported by the IOU community) conflicts with the regulatory duty to focus on system integrity and planning. The elimination of rate freezes have resulted in huge potential price increases. Given today's market model, and more critically, the technological improvements and investments needed for the future of the nation, electricity can no longer be a declining cost commodity. While the industry's 70-year history of driving costs down (supported in a regulatory partnership) is commendable, the current pressures to keep costs flat, protect consumers from the realities of the marketplace, and refuse to incent investment is pushing the sector in the wrong direction just a demands for a more sophisticated and reliable system are increasing.

- Even knowing that capacity will be a problem, there is an underlying level of hesitation in the investment of new coal and nuclear generation technology. This is occurring despite the announcement of 150 new pulverized coal power plants—many of which will not be built due to environmental challenges. It is interesting to note that in a late 2006 survey by Capgemini and the Edison Electric Institute, chief executive officers said they will be spending the most money in the next five years on generation (52.8%) followed by transmission (38.9%) followed by distribution (8.3%). The market reality, however, is that very few of the 150 plants now on the drawing board will ever be built.

- Pressure from Wall Street for double-digit returns from a business that cannot by its very nature provide those returns has forced ill-advised decisions among participants that have damaged their financial status, hurt stock prices, raised the cost of borrowing, and ultimately, sent the sector down a misguided path.

- Not immune from these challenges, public power continues to be concerned about nationwide rules (such as a renewable portfolio standard) and faces significant investments in infrastructure as it builds for the future.

- Politicians are playing to an uninformed public that believes that prices for anything deregulated move only in one direction, and that is downward.

- Significant investments are finally being made in new generation technologies by others, while the current players underinvest in R&D to support the existing thermal fleet.

These observations should not be taken as an indictment of all companies, players, nor regulators, but point to realities in the operation of many companies, jurisdictions, and industry bodies. As the *Wall Street Journal* opined after the 2003 Midwest/Northeast Blackout, there is certainly "enough blame to go around" and there should be a shared sense of failure—but even more importantly, a shared sense of hope as the industry moves to a new transformed destiny. The real challenge lies in developing a logical path to stability and corporate (and system) health while taking into account the regional and technical divergences and allowing for a customer focused business.

The solution clearly must cut through the current inaction and market confusion and will require perhaps an unheralded amount of cooperation. Unless we develop a path for forward motion, the current side-slipping will create unintended consequences for all parties:

- IOUs will be unable to provide rates of return acceptable to the market and face a liquidity crisis that would manifest itself in bankruptcies and high consumer rates—as well as damage to public financial markets.

- Public power, including rural cooperatives, will pay significantly increased transmission charges and assume greater and greater debt loads as the market moves against this sector, which will lead to higher "member" prices.

- Municipal systems will face reduced tax revenues as the market for their supply is impacted and pay higher prices for power at critical times.

- Merchant power will continue to encounter challenging capital conditions which will increase the volatility and amplitude of supply/price swings.

- The federal system will require a significantly higher cost of operations, increased building burden at a time of deficit spending.

- Finally, regulators may be perceived as ineffective and unable to help in any way other than to raise prices for customers.

Ironically, the unintended consequence of these issues will be to drive classes of customers that pay a significantly high share of system costs away from the system and toward distributed generation. This in turn can lead to poorer system utilization, increased air quality problems, and further stresses on the already weakened distribution and transmission infrastructure.

A Vision for the Industry's Future

In its *Electricity Sector Framework for the Future*, EPRI outlined a vision based on input from a variety of key stakeholders. The opening volume of that work describes the coming decades as a new future ripe with business opportunities:

"In broad strokes, the vision is one of a highly reliable, affordable, environmentally friendly power system that provides essential public services and supports the economic aspirations of all classes of customers. It embraces regional and ownership diversity and supports an economic framework of efficient, transparent electricity markets. Within this transformed system, the electricity sector is encouraged and able to invest in new functional capabilities to ensure its operational effectiveness and support the evolving needs of the U.S. economy and society. The key issues limiting the achievement of this transformation, as expressed by the stakeholders, are the depressed financial health of the sector and the continuing jurisdictional confusion in regulation."

This new future will look dramatically different than either the current reality or a simple extension of today's business model. It can open a host of opportunities, build a significantly stronger U.S. economy and fuel a range of new technology opportunities.

The paradigm shift will occur across a range of the broader sector. Customers will be engaged actively in the purchasing and use decisions of electricity and not just be passive, non-involved buyers of bulk commodity. This is not to say that having a commodity business for the provider is a bad thing—it ultimately provides the cash flow and profits for continued investment. Rather this new vision allows for the customer management of electricity (or paying aggregators who will be the proxy for consumers) as the

Benefits of Energy Business Transformation

[Chart showing $B/Year (log scale from 10 to 10,000) vs Years (0 to 30), with two stacked regions: "STEMMING POWER DISTURBANCE COSTS" (lower) and "TAKING THE BRAKES OFF ECONOMIC GROWTH" (upper).]

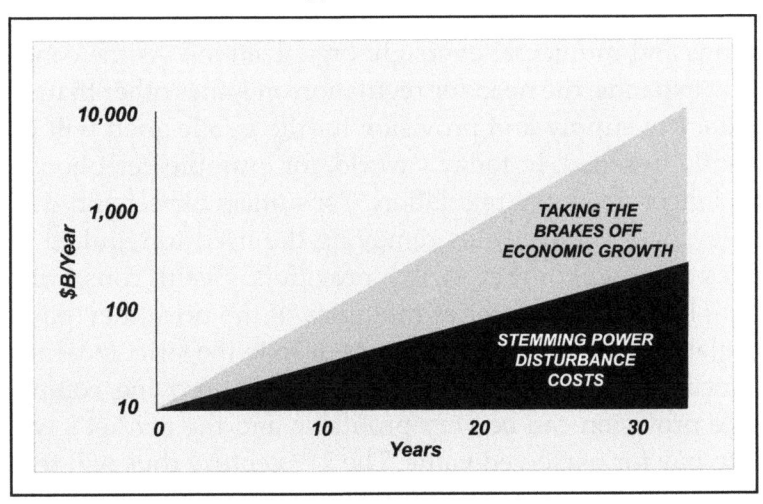

Figure 8-1. The 21st century transformation will be critically important to boosting productivity growth rates, and enabling trillions of dollars of additional revenue for use by both the private and public sectors. This figure shows the combined potential benefits enabled by a transformed electricity infrastructure, both in terms of stemming power-disturbance costs and taking the brakes off economic growth. Given the innovative opportunities that are emerging, it is likely that this projection only scratches the surface of what will be enabled for our economy and quality of life. *Source: EPRI* Electricity Sector Framework for the Future, Volume I, August, 2003

key driver in the business. Just as consumers have taken control in virtually every new technology market, they will drive the electricity industry in ways never envisioned.

Meeting this consumer demand will be a new set of technologies as well as suppliers, vendors, aggregators and marketers who take advantage of new tools to provide yet unimagined services and products. Mobile phone infrastructure has become the conduit for digital photography. Consumers search the web, accessing the internet on their own terms. They demand choices and expect to take this ability into the electric system.

This shift will also create a new role for regulators at all levels from the Federal Energy Regulatory Commission to state commissions and municipal oversight organizations. As the consumer market expands, the need for regulation on issues other than siting, adequacy of supply and provision for the public good will be significantly lessened. In today's world, for example, telephony prices no longer need any regulation. Consumers expect and demand high levels of service (thus removing the need to regulate things such as coverage) or they switch providers. Health conscious consumers buy bottled water at multiples of the price they pay from a regulated municipal water system despite the strict federal, state and local guidelines on water quality—also proving commodity service provision can be very profitable and the market's willingness to pay for perceived value. The 21st century thus will result in regulation being focused on protecting markets and ensuring their operations are open and free from tampering. The use of regulation to force markets to work should also evolve as those markets begin to take form.

This 21st century transformation will mean that all forms of electricity supply, including efficiency, can be effectively monetized. As discussed in Chapter 7, the challenge of the current business model in balancing supply and demand-side resources is in the issue of controllable dispatch and/or storage. The reinventing and reinvestment in the electricity transport system can allow for the effective management of all resources. Controlling energy use for profit or redistribution (As discussed in Chapter 7, The Destiny of Customer Engagement—A Megatrend of Choice) will allow for the market to be managed more effectively, create a working model that builds financial instruments to protect these supplies and put them on a balanced footing. The ancillary benefit will be a greening of the electricity supply, the right financial incentives for cleaner power plants and a reduced dependence on foreign supplies for primary fuels.

As the system is deployed in the 21st century, renewable energy can become a real factor in the nation's fuel mix without the need to artificially stimulate or mandate its use. The lack of predictable load, for example, limits solar to a slice of its potential; wind is gated by the impact of its physical fluctuations on the synchronicity of the electric system; the lack of plug-and-play coupled with the lack of dispatch and control relegates distributed energy resources, at best, to a power quality or peak load offset backup role. In order for the dreams of these alternative fuels and sources to be realized, the system has to be smart enough, fast enough, robust enough and flexible enough to meet the requirements of these newer technologies. Even now, new business models are being developed and refined that will allow wind to be backed by coal; solar to be supported by demand response and intelligent appliances providing flexibility for distribution system operations.

The nation learned all too sadly on September 11, 2001 that system security is all too critical and all too fragile. In the 21st century transformation, security will be matched with reliability in a manner than allows it to be self-correcting, self monitoring and self-optimizing. As the lessons of the internet have taught us, system security and fluidity are built by the dispersion of resources, not the centralization of assets. Despite the widely publicized security flaws created by worms, viruses and other ill-intended attacks, the internet works in an amazingly secure, smooth and transparent manner, transporting billions of pieces of sensitive information every day. This is not to say that there doesn't need to be massive investments in assets—there must be to reach this new vision—but that the assets will be varied, ownership dispersed and management handled in unique and different ways.

Moving into the 21st century, we are already seeing the vision for minimalist environmental compliance (a strategy of some) giving way to the recognition that the electricity business must take accountability for its condition lest regulation and a hostile public will push for measures that are considered intolerable. The deployment of technologies and systems that place the environment at the center of a strong operating business and economy are gaining favor—even if the reality of use trails public pronouncements. The cry for "clean coal" is an example as much hyperbole is used to describe the future, yet when it comes to the actual investments, companies are frightened by the cost and are delaying projects. The irony is that by waiting, the nation's capacity drops and places the business at the mercy of natural gas.

The failure of the current model to monetize the environment, the lack of clarity of regulation now and in the future, the moving battleground of the "bad-pollutant du jour" has handcuffed the current suppliers. The future lies in properly balancing the needs of society, putting a value on all sources of electricity and allowing the system to work effectively. The 21st century, driven by technology and consumers, will properly balance these needs, bolstered by financial markets acting in support of the environment because of monetary incentives. Interestingly, in today's market, the financial incentives have a perverse effect on the environment allowing older plants that have been fully depreciated (and thus showing a lower total cost of operation) to continue service and newly built plants to be idled. This is not done out of mal-intent by the current industry, but rather through financial logic of the current paradigm.

Clean coal and new nuclear are at the core of a solution for the future, yet are perceived as too risky: nuclear due to the lack of disposal and IGCC due to uncertainty about final price and performance. Yet, by not building such facilities, market risk actually rises, with supplies and adequate future capacity in question and a heavier reliance on natural gas.

The attraction of capital in the 21st century will change dramatically to meet these new capabilities. Just as money has flowed into new technology arenas such as telephony, broadband and the internet, the modernization and expansion of the electricity enterprise, along with concurrent consumer demand, regulatory refocusing and an efficiency/environmental marriage, will attract capital investment, technology expansion and a sustainable market sector. This will ironically be coupled with the huge demographic shift over the next 10-20 years as the baby boom generation looks to companies providing solid dividends and returns and potentially away from growth stocks. This demographic group will also demand more of its systems, expect cleaner skies, less environmental impact, more choices and need more electricity to power an expanding, active lifestyle over decades—customers participating in their electricity choice as never before.

In spite of the regional differences in market development and the timing of technological advances, the U.S. electricity sector is expected to eventually reach the 21st century transformation as a matter of necessity, if not survival. The key questions are how long will it take, whether it will it be driven predominately by the current participants or others, and what can be done to enable a smooth and predictable transition rather than a series of disruptive and expensive crisis-laden experiences. Equally important is ensuring that the costs of the transition—and any discomfort experienced by the consuming (and voting) public while reaching the desired state—do not outweigh the benefits; in short, that the transition do no harm.

Former President Bill Clinton held an optimistic view of the electricity business as he expressed in a speech last year in Washington, D.C.:

"What will be the big job generator of the next decade? ...What is our information technology answer to this new decade? My favorite is energy... There is by common consent a trillion dollar untapped market for clean energy and conservation technologies..."

Transformative Paths in Telecommunications

As discussed in detail in Chapter 3, other industries have offered guidance through the transformative process. After intensive infrastructure investment, each of these industries experienced significant evolution—changes that could not have been predicted, regulated or legislated into existence. Rather, the changes occurred as the availability of new technology spurred market demand. A closer look at the unrolling two industry revolutions, that of cable TV and telephony, reinforce the concept that change cannot be regulated, and suggest that the transformation in the electricity enterprise might follow a similar path.

The Cable TV Revolution

Cable television began life as an additional locally regulated monopoly designed to provide better picture quality and a few additional channels in an era where competitors were offering their service for "free." None of the parties involved imagined the revolution that cable television would unleash as the technology progressed to include huge quantities of programming. Consumers had no concept that their lives would be transformed, incumbents in existing programming never envisioned their worlds being turned upside down, and regulators could not anticipate the expansion around their power structure. Even the cable companies themselves did not foresee the advent of the internet (using cable) as a two-way communications venue.

Indeed, it would have been impossible to create scenarios to account for the movement forward in the cable industry—especially if those scenarios had been painted as "choices"—or to predict the timing of the changes, which varied based on locality, region, regulation, and financing. The industry experienced the "wild and woolly" days of no regulation, followed by regulatory certainty (exclusive monopolies for municipalities). It then faced opening of the market by the federal regulators, followed by consolidation and competition from unlikely (dish) companies. Incumbents suffered, new players grabbed markets and market share, many original cable pioneers disappeared, and consumers were charged increasing rates for features they hadn't anticipated needing. Through it all, the industry moved to a transformed (albeit continually changing) marketplace.

In short, the telecommunications industry made change inevitable—if unpredictable—by investing heavily in cable infrastructure. Technology, combined with market needs, drove the changes by regulation, legislation, and litigation, which were supported by the financial markets. While exogenous factors influenced the marketplace and nudged it toward a different path, they could not stop the movement towards a transformed cable industry. Wars propelled programming changes (the CNN phenomenon and 24-hour news); economic downturns crippled heavily leveraged participants but made them ripe for acquisition; and the internet and need for broadband forced the entire sector to reinvest in its infrastructure years before it had been fully amortized.

The Telephony Revolution

The telephony revolution has followed a similar path stemming from, but certainly not bounded by, Judge Greene's 1978 order to break up AT&T. While competitive forces arguably accelerated technology deployment, the industry's current transformed state was made inevitable not by legislation, regulation, or litigation, but rather by market demands sated though the availability of consumer-based technology.

The various periods of confusion in telephony (including the current one) could not have been envisioned as possible end-state scenarios, and market dynamics prevailed despite attempts to control them. For example, limited attempts to re-regulate telephony were followed by litigation that further freed the market, and opening the market to competition has resulted in minimum support standards. Incumbents trying to hold onto infrastructure investments have been thwarted both by regulation and the movement of technology forward.

Throughout these periods of change, consumers, while complaining of the rising costs, have significantly increased the percentage of income spent on telephony. Exercising choice, consumers have accessed a huge variety of features and services that were unforeseen, unimagined, and unknown at the time of initial scenario development. From the introduction of the wall jack as the main portal to the internet to games on cell phones, no amount of foresight, scenario prediction, or legislation could have altered the ultimate course of the telephony business.

Getting There from Here

When asked by tourists for directions to an out-of-the-way location in rural Vermont, locals often resort to the running joke, "You can't get there from here"—meaning the path to the desired point is not obvious and requires travel over a host of interconnected but not necessarily logically laid out roads. A similar analogy can be applied to the electricity business. To take advantage of the technological and business opportunities, it may be that the nation "can't get there from here." The technological, business, and regulatory model needs to be altered to make possible the journey to a transformed future that will benefit all stakeholders—particularly society as a whole.

The future of the industry can be unbelievably bright—or dimmed by a refusal to move forward. While there is certainly no single pathway to that future, it will undoubtedly occur and involve a digital revolution—whether led by the incumbent suppliers or not. The struggle, therefore, should not be about holding back the technological change, since no industry has been able to master that type of control, but rather of controlling the destiny in a way that incorporates disruption, incents innovation, allows for a rate return that is acceptable to markets, and builds a healthy balance of return back into the electricity business

Whether the vision guiding the industry is one outlined by EPRI and industry stakeholders or a version thereof, there is no doubt that the change will come. The only doubt is who will take the advantage, how the public will benefit, which technologies will be employed, how the environment is protected, and how involved the customer is in the equation.

A Healthy Balance

The industry clearly stands at an inflection point resulting from more than 10 years of trying to anticipate market changes in the face of uncertainty. In many cases, that uncertainty led to decisions that focused on the potential of other businesses and away from the core competency of a very successful industry. Chasing returns in non-core venues caused much of the business to look outside of a skill set honed over decades of experience. In the search for the new Holy Grail, much of the industry's "self" was lost.

Given the perilously low level of current infrastructure investment (as illustrated in Chapter 1, Figure 1-1), it's clear the industry will have to spend significant dollars to return the system to health. The only questions that remain concern the timing and pace of the investment, as well as what the industry will choose to invest in. We examine here the costs and benefits of different answers to those questions. Looking to the past as well as to other industries, we summarize what is needed to build the balanced business strategy presented in earlier chapters.

Janus, the Roman god of gates and doorways, otherwise seen as the god of beginnings was shown to have two faces—one which looks behind, and one that looks forward, as through a gate or doorway. Janus presents an excellent representation of our vantage point today in the electricity business (Figure 8-2). Rarely in history have we had the chance to stand at such a turning point and recognize the opportunity we now have of looking back at the lessons learned and looking forward at the possibilities—one of a continuation down a path paved with old and outmoded ways, old and outmoded technology or alternatively, one of tremendous possibilities for transformation of the electricity enterprise and the huge positive impact this can have on our society.

Repair, Replace, Rebuild vs. Investing in Innovation

From a technology perspective, the industry faces two choices: repair, replacement, or rebuilding of the status quo system or taking a bolder leap into developing a new generation of technology. The first path appears to be less risky than the other. But is it? Given the pace and impact of technology, merely taking the easier route of

The Janus Conundrum:
Will We Look Back or Will We Look Forward?

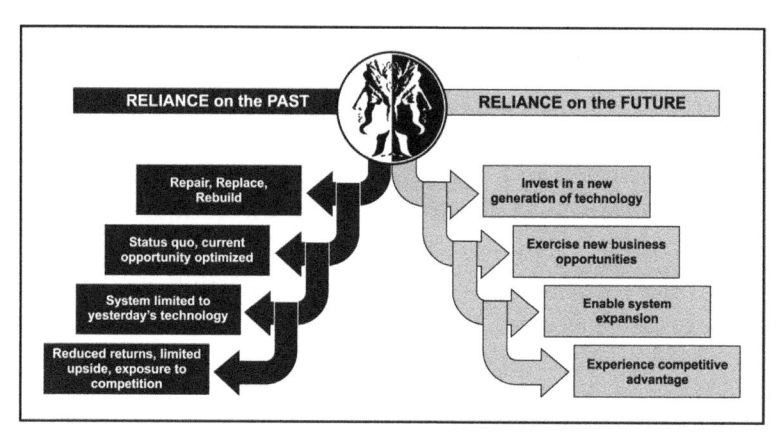

Figure 8-2. Janus, looking both backward and forward is symbolic of our vantage point today—will we walk down a path of business as usual, technology as usual, and a patchwork of solutions, or will we walk down a path of transformation both in terms of business and technology supporting a forward-to-fundamentals approach to electricity's critical role in providing for societal health, prosperity and quality of life?

maintaining status quo technology will potentially lead to a higher level of risk than pursuing new technology.

The major steps in the first path do create near-term advantages over the alternative. Repairing the existing system involves lower initial cost. Part-for-part replacement allows for greater safety in decision making. Rebuilding permits more rapid deployment of the technology. Yet there is also peril in each step. Repairing a system does not allow for new business opportunities and efficiency improvements. Replacement institutionalizes legacy systems and blocks innovation. Rebuilding does not create a robust environment for expanded capabilities.

Locking in technology also creates openings for competition. Businesses adopting the repair, replace and rebuild mentality are at risk of missing marketing opportunities as other players develop technology alternatives and act on related market strategies.

The Destination: A Transformed Electricity Sector

The ongoing expansion of California's notorious "Path 15" discussed in Chapter 6 presents a classic example. Plans for increasing Path 15 throughput from the previous 3,900 MW to 5,400 MW using newly manufactured 1960s equipment, rather than state-of-the-art and state-of-the-possible technologies. In effect, the upgrade institutionalizes system inefficiency.

The option of placing new, advanced technologies and gaining a reward is very real. Implementing and investing in the most advanced power system in the world, complete with flexible AC transmission system (FACTS) would have required an additional $30 million in construction costs. With congestion costing nearly $112 million in 2002 alone, the advanced system would have provided a payback in less than 18 months. Besides alleviating congestion costs, adding a FACTs device on Path 15 could create opportunity and increased revenue for the investing company, while ensuring a better, more robust system.

There is clearly a decision to be made—adding infrastructure within the regulated business or taking the risk on the outside—and the payoff date is uncertain. However, it will take both financial and political wherewithal to make it happen.

The Peril of Industry Hesitations

As discussed earlier in this book, the electricity industry spent much of the mid to late 1990s searching for a customer "call waiting" equivalent—the product or service that would revolutionize and revitalize the electric sector—a quest that resulted in a number of disappointing ventures. As we have argued, the question may have been asked in reverse. That is, the core of tomorrow's promise will lie not in trying to fit new services and products into an existing infrastructure and hoping for profitability, but in investing in infrastructure to support new potentials.

Waiting to invest places companies at peril. Advanced metering, for example, holds the promise of revolutionizing the interactions of customers and the system. In the nearly 15 years since advance meters were introduced, the industry has taken a go slow approach. Even today, driven by EPACT and the needs of lowering operating costs, many companies are looking at ways to avoid replacing meters and installing advanced systems. The good news is that forward thinking firms are moving ahead and installing advanced metering infrastructure, which links advanced, two way meters with meter data management systems to bring a host of knowledge to system operations. By providing the missing connection to the customer, the advanced meter and its successors will open the door to the technological advancements and service opportunities the industry has so desperately sought.

Yet, overall, the industry has held itself hostage in search of the "perfect" advanced metering device that can be cost-justified on the basis of eliminating meter readers. Interestingly, even some of the new installations are not truly with advanced systems meaning 10 years from now those, too, will have to be replaced. Had the phone companies followed this line of thinking, they would have tried to justify implementing advanced technology and switching for the sole purpose of eliminating operators. In fact, eliminating operators may have been one outcome but was not the reason for this change. Another parallel would be waiting to purchase a computer until the cost matches that of a typewriter, foregoing the real and perceived advantages of the new technology.

It may be effectively argued that the new meter, or portal, will pay for itself the very first time it is used for dispatch and load control. A business model can be constructed to justify giving away portal technology at no cost—reminiscent of cell phone, razors, and Pez dispensers—in exchange for the customer relationship and freedom to load manage a certain customer class.

Building a Model for the Future

From many different perspectives, this book has demonstrated the need to invest in advanced technology—not just to maintain the stability of the electricity industry, but to ensure the future health and security of the U.S. economy. The means for such investment would be powered by a business model that strives for balance in several essential areas, including:

- Going "forward to fundamentals" and not just back to basics in re-establishing the core Plain Vanilla business that drives revenues. This includes being "best in class"—establishing best service, best regulatory relations, best customer relations, and best public service image.

- Focusing significant R&D resources on mid- and long-term technology and process development that will strategically build on the company's core business, balanced with infrastructure upgrades that provide an assured rate of return.

- Taking a balanced amount of calculated risk commensurate with rewards in areas where the marketplace has a need for combined service and technology.

- Considering mergers and acquisitions and company spin-offs only when such strategies contribute to the balance of investment in core commodity and service, technology investment and strategic risk

- Searching for adjacencies in business.

- Being prepared, and not hesitant, of new partnerships and business models

- Recognizing the demands of customers will need to be answered proactively

- Insisting on a competitive regulatory model that allows fro new investments, shared risks and shared reward

No matter which path they may choose to take, energy providers need to develop an understanding of the megatrend impacts in order to develop a strategic direction. Recognizing the megatrends and the impact the relevant drivers will have based on the company's makeup, demographics and regional differences, will determine the correct path for each firm.

The demands on the industry have grown to a point that taking no action is, in itself, an action that will ultimately have fatal corporate consequences. Ignoring the destinies does not make them go away. Under investing in infrastructure does not mean an immediate system collapse, failing to build new IGCC facilities does not cause an immediate generation shortage, failing to listen to customers will not immediately result in customer defection, failure to recognize business evolution will not immediately lead to financial instability and being inactive in replacing workers will not mean an immediate problem. However, in the mid to longer term, those who deny or refuse to act on the megatrends will find themselves challenged to provide highly reliable power, to meet environmental constraints, keep up with customer demands, be prepared for competitive threats and to have sufficient human capital to meet system needs.

A critical element is to map current competencies to future needs to understand the gap and prepare to fill it proactively. As some companies are finding out too late, the cost of waiting only adds to the challenge. The successful company will focus clearly on resource needs to meet and lead the market whether it is a small rural electric cooperative or the largest IOU. The impacts of the megatrends will continue to fall across all sectors in the business. Although it is often difficult to move beyond the daily demands of keeping the lights on, companies who understand the destinies and participate actively in their direction will ultimately control their own destinies.

Lessons from the Past

Clearly, Mr. Insull had some of it right. Technology investment allowed his empire to flourish while providing millions of people with electricity at lower and lower cost, enabling the all-electric kitchen, pioneering public transportation, and improving the overall quality of life. Through it all, he relied on technology as a key element to enable new risk strategies that reinvented his business again and again over 35 years.

Although Insull was stripped of his holdings and distanced from his political supporters (Figure 8-3), his assets remained significantly tied up with his holding company, Middle West. In the end, his companies as a whole proved stronger than many other

Figure 8-3. Samuel Insull in 1933 leaving the hospital in Athens, Greece, to face a second extradition demand. Despite the claims made against Insull, his operating companies weathered the financial strain of the depression quite well—twenty-five years after his death, his companies were producing one-eighth of all U.S. electric power and gas and providing these commodities at a cost much less than the national average.
Source: *Chicago Historical Society (International News Photos, Inc.; Negative #DN-ICHi-36570)*

American businesses, which suffered a 40% loss in securities during the depression. Of all the outstanding securities once controlled by Insull, only 20% had forfeited in any way. None of Insull's operating companies had gone into receivership or bankruptcy, and people who had bought their securities lost only a fraction of 1% of their investment. A quarter century after Insull's death, his companies were still producing one-eighth of the electric power and gas consumed in the United States which were still selling for much less than the national average.

A healthy balance of commodity, risk, and technology development remains the core strategy for electric utilities to successfully overcome the business and technical hurdles facing them today. Once energy companies create a business model that feeds this healthy balance, encourages a constant reinvention, and develops a reward structure for rate of return, the industry will be on the road to a solid and sustainable future.

Resources
"Electricity Sector Framework for the Future, Volume I," EPRI Report August, 2003
"New Economy Focuses on Energy Ideas, Ken Silverstein," Utilipoint Issue Alert, March 2004
National Oceanic and Atmospheric Administration
Insull, Forrest McDonald, University of Chicago Press, 1962
"Pulse Survey," EEI International Utility Conference, May, 2007

Conclusion: Ten Ideas for the Future

Looking to the future, we conclude with some ideas to initiate the debate as the industry endeavors to transform itself to maintain its century-long legacy as the crucial underpinning of our nation's robust economy. Energy companies must work with government, industry, and other stakeholders to create the vision and reality of a strong and sustainable energy sector. This collaborative effort will develop solutions and strategies that will transform our industry, as well as with the society that depends on it.

Below are some ideas intended to initiate this process. As the transformative work moves forward, stakeholders will add to and delete from this list—sparking rich debate and fueling actions that together will generate the foundation for a solid, robust industry capable of not only sustaining our economy—but helping the world of the 21st century resolve some of its most pressing and urgent quality of life issues.

1. *Stop waiting for the perfect solution and perfect technology.* In this engineering and financially based industry, companies too often seek the perfect solution rather than deploying the best available technology. As noted, the metering business is a primary example of delaying deployment of good technology while searching for the ideal solution. Clearly, the inherent risks of moving on a less-than-perfect technology need to be considered before initiating a strategy. However, such risks should not be the key factor in the go/no go decision.

2. *Ensure that the business decisions focus on building your new commodity or service annuity, even at the risk of hurting the old.* High-technology companies have learned this lesson the

hard way by holding on to technologies too long and propping up business models past their prime. All businesses need mainstays of profitability, but should be willing to put them at risk when pursuing new options.

3. *Invest in infrastructure with no regrets: it may be costly in the short run but will pay off in the long run.* In a pure rate of return, cost of service model, justifying infrastructure investment, while locked in a business practice as still followed today, is potentially detrimental to the companies and the public. The failure to invest substantially decreases future potential, raises the risk of being left behind in the marketplace, and ensures lower earnings and higher operating costs.

4. *Technology is the enabler.* If your firm does not embrace it, your competitors will to your disadvantage. The electricity industry has been and is still insulated from a number of market forces. The electric system cannot be outsourced overseas and cannot be reliably substituted. Yet those market forces will be unleashed either by deregulation or by the market bypassing the current system. The search for power quality and reliability, for example, may drive industrial and commercial customers to new protection schemes and vendors for increasingly sensitive equipment. The leap to turning to the same schemes and vendors for primary power is not very far.

5. *Begin speaking to your customers about your business in ways your mother would understand.* Like many industries, electricity has shrouded itself in complex concepts and technologies. This protectionist view has created some of the challenges faced by the business today—including customer assumptions that electricity comes from the outlet with no supporting systems; that renewables can simply replace fossil fuels; that the business refuses to control some pollutants simply because of costs rather than because of technological gaps; and that electricity is controlled by regulation and legislation and not by the laws of physics.

6. ***Best in class will beat cheapest in class any day.*** Pursuing the cheapest rather than best, highest earnings rather than a stable return, and reduced infrastructure investments and operating costs promises a cloudy and expensive future. For a solid future, regardless of whether or not deregulation continues, following a best practices approach that targets achieving the highest level of service and maximum operating efficiency will be the most promising course.

7. ***Excellent regulatory relations pays dividends.*** Every player in the market has a regulator of one form or another. For those regulated businesses, make the regulator your partner. The era of conflict between regulators and energy companies will draw to a close as customers become empowered and regulation focuses on protecting markets during this new wave of competitive regulation. Competitive regulation will require cooperation to be successful. While many of the current jurisdictional issues are managed through an adversarial proceeding, the future lies in a different path.

8. ***Not taking action in re-envisioning your future is taking action—in the wrong direction. Markets in pursuit of a new future will bypass companies standing still.*** Technology does not remain static, and customers will demand greater levels and differing levels of service. Opportunities lie in providing those new business opportunities and models, and failure to act is in effect making a decision. Hotels, for example, failed to see their phone business wiped out by cell phones. Instead of moving ahead with other options, they have simply raised the price to astronomical levels—which further drives customers to alternatives.

9. ***Adjacency will beat synergy any time—especially companies seeking synergy for savings, and not to reinforce the commodity business.*** The industry abounds with examples of acquisitions driven by perceived operating savings rather than by

new business opportunities. Many of these have failed to meet the expectations of regulators, the market, and certainly the public. The adjacency model of buying/building business that allow for expanded opportunity will be the winner in the long run.

10. *There is no business more critical, in peril, and yet offering greater opportunity than the electricity sector.* All of the business forces point to a dramatic shift in the functioning, profitability, reliability, and sustenance of the current business model for the electric industry. The model which was so robust in the past is under severe pressure—felt by public power, federal power systems, and investor owned utilities alike. Competition, lack of investment, pressures on earnings, and a public with increasing demands will force the change.

Index

Symbols
2003 Northeast/Midwest Blackout 175
2003 Tax Reform Act 12
2005 Energy Policy Act (EPACT) 81

A
advanced metering 37, 87, 196, 216
   infrastructure 69, 181
   systems 104
age of utility workers 76
aging workforce 80
Allegheny Energy, Inc. 159
Ameren 37
American Electric Power 128, 130, 155, 193
AMI 184
Apple Computer 44, 61
Aquila Energy 28
Arizona Public Service 130
Austin Energy 214
automated meter 169
   reading 107

B
Babcock and Brown Infrastructure 43
Bechtel Corp. 128
biomass 144
boomers 71, 72, 75, 84, 92
BP 137

broadband internet 26
broadband over power lines 209

C
California Energy Commission 209
California power crisis 189
call waiting 26, 34, 59, 222
   model 25
cap-and-trade 157
capital expenditures 6
Caprock Holdings 43
carbon capture 110, 117, 165
carbon conflict 195
carbon constraints 110
carbon credits 73
carbon sequestration 129
carbon tax and/or trading 36
clean coal 110, 165, 231
climate change 215
coal 124
coal-fired power 119
communications 63
competitive deregulation 64
competitive regulation 38
   investment model 39
Comverge 213
ConocoPhillips 129
consumption 101
Control 4 217
customer choice 2, 67
customer empowerment 206

customer engagement 184

D
demand 95
demand response 101, 102, 112, 164, 208, 209, 215, 219, 231
demand side 171
    management 21
demographics 71, 91, 105
Department of Energy 170
deregulation 19, 21, 63, 75, 202, 204, 223
digital devices 175
distributed energy resources (DER) 14, 114, 148
distributed generators 213
distributed power 114
distribution 185
    automation 104, 186, 196
    management systems 197
    upgrades 179
DSM 162
DTE 42, 45
Duke Energy 128
dynamic thermal circuit rating (DTCR) 188

E
ecological assets 159
electric power reliability 176
Electric Power Research Institute (EPRI) 119, 224
enabling 187
energy/information portal 212
energy consumption 96
energy demand 96
energy efficiency 112, 204, 208, 215
Energy Policy Act of 2005 134, 171, 204

energy portal 207, 216
energy service companies (ESCOs) 26, 52
energy storage 152
Enernoc 213
Enron 14, 43, 54, 224
environmental controls 75
EPA 162
EPACT 24, 43, 45, 161
EPACT 2005 224
EPRI 236
ESCO 53
Exelon Nuclear Fleet 114

F
FACTS 186, 198
fault anticipation 182
fault anticipators 188
Federal Energy Regulatory Commission 205, 230
FERC 205, 230
forecasting 182
fuel cells 68, 149, 150
fuel diversity 116
FutureGen Alliance 127

G
GE 137
General Electric 32, 83, 128, 129, 137
geothermal 145
global warming 112, 134
Golden Valley Electric Association 155
green 105
    movement 170
    power 73, 94
greening 230

Index

H
Hawaiian Electric Company 139, 141
hydrogen 147, 194
    economy 68
hydropower 146

I
IGCC 87, 117, 125, 131, 166, 232
independent system operator (ISO) 24
infrastructure 75, 93
    aging 34
Insull, Samuel 7, 9, 50, 243
intelligent grid 169
intelligent infrastructure 220
intelligent networks 170

K
KCP&L 37
KKR 42, 45
knowledge 90
    gap 78
    management 85
Kohlberg Kravis Roberts (KKR) 43

L
linemen 81
liquefied natural gas 133, 165
LNG 133, 165
load control 104, 164
Long Island Power Authority (LIPA) 192

M
M&A 29, 67, 72
megatrends 15, 16, 68
mergers and acquisitions 29, 67, 72
meters 81
meter readers 82
MidAmerican Energy 42
migration 92

N
natural gas 115, 132, 133, 172
negawatt 20
NERC 24
New Century Energies 78
new plant construction 113
NiSource 78
northeast power outage 63
NRG 115
    Energy 130
nuclear 232
    energy 134
    power 83
    units 82

O
outage management 169, 182
outsourcing 77, 79, 88

P
PacifiCorp 42
Path 15 189
Peabody Energy 128
peak demand 210
PG&E 28
plant construction, new 113
PNM Resources 43
portal 209
Portland General Electric 43
power delivery infrastructure 81
power demand 202
power interruption 178
power plant construction 109
power quality 73, 94, 96, 102, 175
power trading operations 25

public good 62
public power employment 77
Public Utilities Holding Company Act 30, 50, 213
Public Utility Regulatory Policies Act (PURPA) 20

Q
qualifying facility 20

R
R&D 57, 214
    investment 6
real-time pricing 206
regulated environment 46
regulated monopoly 9
regulation 10, 218
renewable energy 114
renewable fuels 137
renewable portfolio standards (RPS) 110, 138, 143
renewable resources 112
resource planning 94
restructuring 62
risk management 102
rural electric co-ops 35

S
SDG&E 143, 172
sensors 187, 195, 196
Shell Corp. 129
Site Controls 213
Smart Grid 40, 87
solar 141
Southern Company 83, 126, 128
Southwire Company 192
SQRA 36, 69, 174, 220

standard market design 22
Starr, Chauncey 194
stranded assets 22
substations 185, 186, 197
superconducting magnetic energy storage 156
superconductivity 194
supervisory control and data acquisition 184
sustainable business model 44

T
telecommunications 59
    deregulation of 59
Tennessee Valley Authority 45
TransAlta 130
transmission 178, 189
    bottlenecks 179
Tucson Electric Power 43
TXU 28, 78, 116

U
utility workers, age of 76

V
venture side 49
virtual power plant 210

W
Wal-Mart 61
Welch, Jack 32
Western Area Power Administration 189
wind 139
workers 71
workforce 74, 78, 86